AF651243

Encyclopaedia of
Animal Cell and Tissue Culture

Encyclopaedia of Animal Cell and Tissue Culture

Volume 2

Editor:
Dr. Anil M. Mane

AGROTECH PRESS
JAIPUR ▪ NEW DELHI

AGROTECH PRESS

Regd. Off. "Bony Residency", Gate No. 2
Opp. Tilak Public School, Vishweswariya Nagar,
Gopalpura Road, Jaipur - 302018, India
Ph: 0141 - 2761280, Telefax: 2761381
E-mail : agrotechpress@gmail.com

Delhi Office: 102, 1st Floor, Satyam House,
4327/3, Ansari Road, Darya Ganj,
New Delhi - 110002, India
Ph: 011-45652440
visit us: www.agrotechpress.com

Distributed By
KEYA INTERNATIONAL PRESS LLP
E-mail: keyainternationalpressllp@gmail.com

Published in India, 2025

© All rights reserved

ISBN : 9789384568276 (3 Vols. Set)

All Rights Reserved. Neither this publication nor any part thereof may be reproduced, stored in a retrieval system, or transmitted, in any form or by any means, electronic, mechanical, photocopying, recording, scanning or otherwise, without the prior written permission of the copyright owner.

Typeset by Thomson Press

Printed and bound in India by Thomson Press Ltd.,
New Delhi

Contents

Preface

Tissue culture is the growth of tissues and/or cells separate from the organism. This is typically facilitated via use of a liquid, semi-solid, or solid growth medium, such as broth or agar. Tissue culture commonly refers to the culture of animal cells and tissues, while the more specific term plant tissue culture is being named for the plants.

In 1885 Wilhelm Roux removed a section of the medullary plate of an embryonic chicken and maintained it in a warm saline solution for several days, establishing the basic principle of tissue culture. In 1907 the zoologist Ross Granville Harrison demonstrated the growth of frog nerve cell processes in a medium of clotted lymph. In 1913, E. Steinhardt, C. Israeli, and R. A. Lambert grew vacciniavirus in fragments of guinea pig corneal tissue. In modern usage, "tissue culture" generally refers to the growth of eukaryotic cells in vitro. It is often used interchangeably with cell culture to specifically describe the in vitro culturing of sperm donor cells. However, "tissue culture" can also be used to refer to the culturing of tissue pieces, i.e. explant culture or whole organs, i.e. organ culture. It is a tool for the study of animal cell biology in vitro model of cell growth to allow a highly selective environment which is easily manipulated. A veterinarian may request that a tissue sample be taken from an animal to assist in the diagnosis of a condition or for research purposes. These tissue samples can be used to determine a contaminant that is causing a pet or animal illness. A tissue culture from a lump or mass can help determine the likelihood that the cause is cancerous or the chances of the condition spreading. Animal tissue culture collection techniques are very similar to human collection methods. Most animal tissue cultures are taken through the technique of blood drawing. This method is completed by inserting a hollow needle into the vein of the animal and attaching a vacuum tube to the needle, drawing the blood out of the vein. This method is slightly painful and can cause an animal to spook or draw back, but it is overall harmless and minimally invasive. The tissue—or blood—that is drawn into the vials can then be tested for viruses and virus antibodies. Bacteria can also be diagnosed through blood work and various illnesses can be diagnosed in part with the changes in blood gas and content levels. Biopsies allow the medical and scientific teams working with the animal to collect a larger sample of tissue for study. This tissue can come from any portion of the animal. The most common method of obtaining a biopsy sample is the punch method. This method removes a small amount of skin or other tissue by creating a small hole in the tissue, collecting what was there previously. This method can be painful, and most animals that undergo a punch-style biopsy are sedated to prevent trauma. A needle biopsy can also be used, which is where the tissue is drawn into a hollow needle and placed in a sterile tube. This method is frequently used if the area being sampled is small or if the tissue is difficult to reach with a biopsy punch. Both biopsy styles can be painful and invasive. Swabbing is a common method of non-invasively collecting animal tissue cultures. This technique involves wetting a swab with a sterile solution and wiping the mouth or skin of the animal with the swab. That swab should then be allowed to dry in a sterile environment and can then be tested for DNA or genome testing. Oral or skin infections can also be detected using this technique of tissue culture collection.

This book covers a formal introduction and provides the basics in understanding the concept of tissue culture traversing through the state of the art technologies such as stem cell, hybridoma technology, transgenesis, cloning and scaling up technologies. This book will impart up to date knowledge and confidence to execute tissue culture related work with ease.

—Editor

1

Conservation through Tissue Culture Techniques

Multiplication through tissue culture techniques will be done for selected trees which are, high active-ingredient yielding, fast growing, and of good form, using indigenous knowledge. The expected outputs from the research are conservation of selected overexploited prunus germplasm from various provenances, through the establishment of gene banks, distribution of already characterized and evaluated, tissue culture multiplied germplasm for planting by farmers and documentation of the results and information for dissemination to stakeholders.

To date successful plantations and enrichment plantings have been limited to Kenya where over a period of 90 years, prunus Africana has been planted by the forest department for timber. However, their planting material is often of unknown origin and of narrow genetic base. They are therefore a limited utility for conservation purposes. A study was carried by Cunningham and Mbenkum, to investigate the economic feasibility of different planting systems enrichment planting, small-scale farming and plantations for prunus as an alternative to the current overexploitation of wild populations in Cameroon.

Some of the recommendations made were that low technology propagators at field sites would boost the number of young seedlings being produced, and research needs to be conducted for selection of fast growing, high active yielding prunus Africana cultivars. If more attention is placed on the origin and genetic diversity of material, enrichment plantings and plantations can serve a useful conservation function and act as a source of planting material for other users. This research therefore aims to select in natural forests and farmer's fields overexploited Prunus germplasm, and plant them in gene banks, for their conservation. Due to the intermediate nature of the seed, propagating the tree through seeds may be difficult. Furthermore, it takes 15 years for the trees to seed. Propagating prunus through tissue culture is a rapid method of multiplying the prunus Africana and is a faster way of achieving our objectives of conservation.

Prunus Africana is popularly known as Pygeum, and is a large evergreen tree that grows in the afromontane of Africa. It is the only species of Genus Prunus native to Africa and can grow to a height of up to 40 meters. It has pendulous branches with thick oblong-shaped, leather-like, mat coloured leaves and creamy white flowers.

The fruit (drupe) resembles a cherry when ripe. Traded internationally and harvested from the wild, the Prunus is hardly a minor forest product. The tree is valued for its medicinal extract from its bark for the treatment of benign prostatic hyperplasia, a common condition in elderly men. Annually, approximately 4000t of bark are harvested primarily from Cameroon, Equatorial Guinea, Kenya and Madagascar to meet this demand.

This bark exploitation has caused serious damage to wild populations of Prunus including trees inside forests of high conservation value, leading to concerns on the long term sustainability of harvesting and conservation of this tree species. Diversity of other uses of this tree include making of handles for axes and hoes, firewood and poles. A serious constraint in the expansion of on farm tree planting is seed availability. Here a key difficulty is the intermediate nature of the seed. Furthermore seed yields fluctuate widely between years. Seed shortage is likely to be exacerbated in future as the size of natural populations diminishes. The approximate time to first flowering and fruiting is 15 years.

Raising Prunus through tissue culture techniques will therefore accelerate their production. The transfer of germplasm from the wild into on farm niches helps to preserve valuable genetic resources particularly if attention is paid to the origin and diversity of cultivated material. In addition, cultivation of endangered forests species takes pressure off their natural resource base, thus promoting the conservation of natural forests.

Objectives

- To gather through explorations and collection expeditions in different ecological zones, overexploited Prunus germplasm as well as meaningful and rich stock of Prunus with superior characteristics.
- Rapid multiplication (through Tissue culture techniques) of both the heavily exploited germplasm and the Prunus with superior characteristics, for establishment of conservation stands and their distribution to farmers.

Hypothesis

Conservation of Prunus Africana can be enhanced through tissue culture techniques.

Expected outputs:

- Conservation of the overexploited Prunus germplasm of different provenances in established conservation stands.
- Continued availability of the selected Prunus germplasm through their continuous multiplication.
- Documented research results for ease of dissemination and retrieval.

Selection

Three sites have been chosen to be areas in which selection of Prunus Africana will be done. These are Mt. Kenya forest, Mt. Elgon forest and the Abardares. These sites were chosen as they are areas in which the tree is known to occur naturally and is heavily overexploited. Random selection of trees will be done in the chosen respective provenances with the help of forest officers and farmers in the communities with indigenous knowledge. Selection will be done for trees that are high ingredient yielding, fast growing, early seeding, and have good form *i.e.* trees that are straight, with no mechanical damage no forking or kinking, no dwafing or diseased.

Germplasm will be collected from trees which have passed selection criteria as well as from trees that have been overexploited and are becoming extinct. Measurements to be taken for the selected trees are diameter at breat height and height of the trees. Other data to be noted on site include, approximate age and location of the tree. Branches or coppices will be collected from 25 randomly chosen trees per provenance using secateurs. They will be packed according to provenances and carefully labeled using marker pens.

The packaging will consist of polythene bags containing water, which will be stored in cooler boxes. Seeds will be collected from 25 randomly chosen and supposedly unrelated individuals using a seed harvester, packed in khaki bags which will also be labeled using a marker pen. The collected coppices will be brought to the lab for propagation through tissue culture techniques. Bark samples will also be collected from the selected trees and packed in Khaki bags, which will also be labeled.

The bark samples will also be taken to selected laboratory for chemical analysis. Collected seeds will be germinated in seed beds and then transplanted into potted tubes containing soil media. They will be planted in conservation stands once they are ready for planting.

Field Trial Design (Conservation Stands)

The sites for planting the propagated material for both seed and tissue culture from the tree provenances will be identified. The land will be cleared, ploughed and planting holes dug. The sites will be at isolation belts of between 300 to 500 meters from other natural vegetation that have Prunus trees.The propagated material ready for field planting, will be planted out in the field during the onset of the rains, using local labour. The material from the three provenances will be planted in each of the provenance. Each provenance will have a total of 533 trees and will be replicated 3 times across the trial design. Each replicate will therefore have 133 trees from each provenance. There will be 4 replicates for each of the provenance. Replicates will be used for planting prunus propagated from seed.

The site will be visited twice a year for monitoring and maintenance *i.e.* weeding. Data will be collected once a year. This includes data for survival, height, root collar, diameter and diameter at breast height for 10 randomly selected trees per plot of 133 trees. The data will be analysed at the nursery. Thinning will also be done when necessary.

Tissue Culture Procedures

The explants from the field will be stored in a refrigerator, while still in the polybags with water to keep them fresh. Tissue culture will be done progressively, one provenance at a time. The explant to be used here are shoot tips from the Prunus.

However, a protocol will have to be worked out through trial and error, to establish which media combination will be successful in the propagation of the prunus. The culture media should have a pH of 5.7 +/- 0.1, before agar is added. This pH is achieved by measuring the media pH using the pH meter and adding HCL or NaOH to adjust the pH accordingly.

Agar is always added last, and it should be mixed well, using heat, preferably in a microwave The prepared media will be discharged into test tubes (5mls per tube) using a syringe. The test tubes will be covered and packed in an autoclavable tray and placed in the autoclave. The autoclave will be set to 15 minutes at slow exhaust. When the autoclave cycle is complete and the pressure has reached zero, and temperature below 100°C, the autoclave door will be opened and the tray with tubes removed.

Surface Disinfection of Explant Material

The explants will be trimmed to size convenient to handle and placed in a container with little quantity of water to maintain the moisture until processing. The water in the container is then discarded. The material is then put into 70 per cent ethanol for 25 minutes, followed by 5-10 per cent sodium hypochlorite solution with few drops of tween for 20 minutes. The material is then rinsed with 3 changes of sterile distilled water.

The explants are then kept in the last change of water. If the material is heavily contaminated, a double disinfection process with sodium hypochlorite may be used. The explants are now ready to take to the transfer chamber. This last step will be performed inside a laminar flow hood.

Starting the Explants (this is done in the Transfer/Innoculation Room)

The transfer chamber should be ready with the walls and workspace sprayed and wiped with 70 per cent ethanol. Our hands should also have been washed with soap and water till the elbows, first, then sprayed with 70 per cent ethanol. There should be a container of 70 per cent ethanol to sterilize and rinse the instruments, *i.e* the blade (no 10) and forceps. Immerse the forceps and scapel for 30 seconds or more in the 70 per cent ehanol then burn off the etanol over the flame from the bunsen burner.

With the forceps place an explant tip on the petri dish. Place the forceps in the other hand to hold the tip while the first hand uses the scalpel to cut off 1 cm of the basipetal end of the shoot tip. Place the scapel in the70 per cent ethanol, move the forceps to the other hand. Grab a sterile test tube of medium with the other hand, hold it by the base. With the little finger of the hand holding the forceps, remove the cap and cradle it there while we use the forceps to firmlty lay the bud on the medium. Recap the test tube and seal with a piece of scotch tape (or Para film).

Physical Components of Tissue Culture Technology

The physical components of a typical plant tissue culture facility include equipment and buildings with preparation room, transfer room, culture or growth room, hardening and weaning area, soil-growing area (greenhouses, plastic tunnels), packaging and shipping area, and related facilities – office and store for chemicals, containers and supplies. Careful planning is the first important step when considering the size and location of a facility.

The size of the physical components of a tissue culture facility will vary according to its functional needs – the volume of production. It is recommended that an existing facility should be visited to view the layout and operational needs before starting a new facility. A number of low-cost alternatives can be used to simplify various operations and reduce the costs in a tissue culture facility.

In designing any laboratory, big or small, certain elements in its design and layout are absolutely essential for its successful operation. Correct design of a laboratory will not only reduce contamination, but also achieve a high efficiency in work performance. Properly planned and designed laboratories can reduce both the operational and energy costs. A tissue culture laboratory must be designed to accommodate the equipment and its use in the various stages of micropropagation in the most efficient manner.

General Consideration for Location

A convenient location for a small laboratory can be a room or part of the basement of a house, a garage, a remodeled office or a room in the header house. The minimum area required for media preparation, transfer and primary growth shelves is about 14 m^2.

Walls should be installed to partition different areas. Before setting up a commercial micropropagation unit, it is essential to check out the area keeping in mind the climate, and access to water, electricity, transportation, and infrastructure for supplies.

A temperate climate is usually better suited to tissue culture ventures. This greatly reduces the cost of cooling required to maintain the temperature for optimum growth of the cultures. The availability of electricity and water is of utmost importance, and should be taken into consideration while choosing the location of the facility. For example in India, of the 76 commercial tissue culture units, nearly 52 are located in and around the cities of Bangalore and Pune, where the climate is moderate.

Hence, However, in cities like Delhi, which have extremes of climate, tissue culture facilities require both heating and cooling. Disruption of power and water supply causes major breakdown in the smooth running of tissue culture units. Poor quality water adds to the cost of media.

Location consideration should also include a check with local authorities about zoning and building permits before construction begins. The buildings should be located away from sources of contamination such as a gravel driveway, parking lot, soil mixing area, shipping dock, pesticide storage, and dust and chemicals from fields. Sanitation of the area is important in selecting location.

The units should be located in areas where insect populations are minimal and air has low dust and pollen counts. In the case of export-oriented units, these should be near an international airport, which will reduce the time lag between packaging and shipment. This is critical to assure timely delivery of quality tissue culture products. For new ventures, the size of the facility should be kept small until the market acceptance is ensured.

Design and Layout of the Laboratories

Plant tissue culture laboratories have specific design requirements. Careful initial planning is, therefore, a prerequisite for successful running of a facility. The location and design of the laboratories should take into account isolation from foot traffic, control of contamination from adjacent rooms, thermostatically controlled heating and cooling, water supply and drains for a sink, adequate electrical service, provisions for a fan and intake blower for ventilation, and good lighting. Large sized facilities are frequently built free standing. Although more expensive to build, the added isolation from adjacent activities keeps the laboratory clean. Prefabricated buildings make convenient low-cost laboratories. They are readily available in various sizes in many countries. Prefabricated buildings assembled on site can also be used. A single span building allows for a flexible arrangement of walls for dividing into convenient sized rooms.

The floor should be of concrete or capable of carrying 170.5 kg/m^2 (50 pounds per sq. foot). Walls and ceiling should be insulated to at least R-15, and covered inside with waterresistant material. Windows, if desired, may be placed wherever convenient in the media preparation and glassware washing rooms. The heating system should be capable of maintaining room temperature at 20°C during the coldest part of winter. A minimum of 2cm pipes should be used for water supply. Connection to a septic system or sanitary sewer should be provided. Air conditioning requirements should carefully estimated. Electrical service capacity for equipment, lights and future expansion should be calculated. For safety reasons, the electrical installation should be carried out professionally. Most electrical wiring will require 220 Volts, and autoclaves 230/250 Volts.

The areas such as the media preparation room, inoculation room and growth chambers should be isolated as 'clean zones'. The office, storage area, staff centre and packaging rooms can be maintained under ordinary conditions. The working areas must be demarcated according to the activities involved in the facility. Cleanliness is the mazor consideration when designing a plant tissue culture laboratory to minimize contamination. A positive pressure module should be installed to circumvent air intake from outside. Routine cleaning and aseptic procedures can decrease contamination losses to less than 1 per cent. An enclosed entrance should precede the laboratories, and sticky mats should be placed to collect dirt from shoes.

The traffic pattern and workflow in the laboratory must be considered to maximize cleanliness. The cleanest rooms or areas are the culture room (aseptic transfer area) and the growth room. There should be no direct access to these rooms from outside. The media preparation area, glassware washing and storage areas should be located away from these rooms. The growth room and aseptic transfer rooms should be fitted with see-through doors and should be adjacent to each other. Traffic through these areas should be minimal and restricted to the personnel working under laminar flow cabinets. Ideally, the media preparation area leads into the sterilization area, which leads into the aseptic transfer room, and eventually to the growth room. Temperature and fire alarms must be connected directly to telephone lines to give fast warnings.

An emergency generator should be available to operate essential equipment during power breakdown.

Essential Equipment

The basic equipment in most tissue culture facilities includes the following:

Autoclave

An autoclave is basically a large-sized but sophisticated pressure cooker, and is used for the sterilization of the medium, glassware and instruments. Autoclaves of different sizes are available commercially. High-pressure heat is needed to sterilize media, water, and glassware. Certain spores from fungi and bacteria are killed only at 121°C and 1.05kg/sq.cm (15 pounds per sq. inch) pressure. Self-generating steam autoclaves are more dependable and faster to operate.

Laminar Airflow Chamber

The laminar flow chambers provide clean filtered air that allows cultures to be handled under contamination-free environment. Several types of laminar flow chambers are sold in the market and are available in different sizes. The laminar-flow cabinets are located in the culture transfer area. Some large-sized laboratories have sterile rooms in addition to laminar flow cabinets.

Other Equipment

The sterilization of instruments, such as the forceps, scalpel holders and blades is achieved with either gas flamed burners or with glass-bead sterilizers. The medium preparation room usually has the following equipment. A refrigerator-freezer to store chemicals and stock solutions, weighing scales for large amounts of over 10 g, and an analytical balance with1 mg accuracy, a magnetic stirrer for the agitation, and a pH meter. Small laboratories may locate the refrigerator under the workbench to save space. High quality balances are essential in a tissue culture laboratory. Most laboratories have top loading balances, which allow quick and efficient weighing. A hot plate with an automatic stirrer is needed to for preparing the media before autoclaving.

The pH meter is needed to determine pH of the media. Some laboratories use pH indicator paper, however this method is considerably less accurate, and can severely affect the results. An aspirator can be attached to a water tap for filter sterilization of chemicals and for surface sterilization of the plant material. However, vacuum pumps are faster and more efficient, but also more expensive. A drying oven is required to keep glassware such as beakers, flasks and cylinders, and is also useful for dry sterilization of scalpels and glassware, such as Petri dishes, pipettes and others. The media containing carbon sources (*e.g.* sugars) and growth regulators are sterilized in the autoclave, but sometimes, aseptic filtration is better to avoid breakdown of heat-labile chemicals. The water still is also located in the medium preparation area. To prepare media, distilled or de-ionised water is generally used, although tap water can be used in some cases.

Optional Equipment

A variety of non-essential equipment is used in tissue culture laboratories. The specific requirements determine what need to be purchased. Microwave ovens are convenient for defrosting stock solutions and pre-heating agar media. Most laboratories have a dissecting microscope to excise small explants. Laboratory glassware washers or regular dishwashers can be used for replacing manual labour. Automatic media dispensers are helpful to pipette pre-set volume of media. A gyratory shaker or a reciprocal shaker is necessary if micropropagation is based on liquid media or suspension cultures. Computers, photocopiers and fax machines are helpful for easy data management and maintenance of records. Some of the equipment may be costly, but goes a long way in saving time and labour and is essential for rapid communication in the competitive world.

Activity Specific Requirements

Based on the different activities of a tissue culture, a facility can be divided into semiclean, clean and ultra-clean areas. The semi-clean areas comprise of the washing room, office and staff restrooms, where there is no need for maintaining sterile conditions. The clean areas encompass the media preparation and sterilisation rooms, which have to be sufficiently clean. High sterility has to be maintained in the culture transfer rooms and the growth rooms, which constitute the ultra-clean areas.

Glassware Washing and Storage Area

The glassware washing area should be located near the sterilization and medium preparation rooms. This area should have at least one large sink but two sinks are preferable. Adequate workspace is required on each sides of the sink; this space is used for glassware soaking and drainage. Plastic netting can be placed on surfaces near the sink to reduce glassware breakage and enhance water drainage. The outlet pipe from the sink should be of PVC to resist damage from acids and alkalis.

Both hot and cold water should be available and the water still and de-ionization unit should be located nearby. The choice of electrical washers should be based on the projected use, durability, reliability and cost, and service availability. In India and some other developing countries, where labour is relatively cheap, washing is done manually. The washing room should be swapped periodically.

Mobile drying racks can be used and lined with cheesecloth to prevent water dripping and loss of small objects. Ovens or hot air-cabinets should be located close to the glassware washing and storage area. Dust-proof cabinets and storage containers should be installed to allow for easy access to glassware. When culture vessels are removed from the growth area, they are often autoclaved to kill contaminants and to soften semi-solid media. It should be possible to move the vessels easily to the washing area. The glassware storage area should be close to the wash area to expedite storage and access for media preparation.

The Store

It is advisable to have a separate area for storage of chemicals, apparatus and equipment. It would not only facilitate constant availability but also save cost from bulk purchase. Chemicals required in small amounts should not be purchased in large quantities as they may lose their activity, pick up moisture or get contaminated. Such problems can be overcome by purchasing small lots on a regular basis.

Media Preparation and Sterilisation Area

The media preparation room should have smooth walls and floors, which enable easy cleaning to maintain a high degree of cleanliness. Minimum number of doors and windows should be provided in this room but within the local fire safety regulations. This reduces cost and contamination. The media preparation and sterilisation can be carried out in the same area but preferably in different rooms, which need not be separated with doors. Media preparation area should be equipped with both tap and purified water. An appropriate system for water purification must be selected and fitted after careful consideration of the cost and quality. A number of electrical appliances are required for media preparation; hence, it is essential to have safety devices like fire extinguisher, fire blanket and a first aid kit in the media preparation room. A variety of glassware, plastic ware and stainless steel apparatus is required for measuring, mixing, and media storage. These should be stored in the cabinets built under the worktables and taken out for use as and when required.

This would save the cost and space for building storage shelves. The use of glassware should be kept at a minimum, as it will help in reducing losses due to breakage. As far as possible, plastic ware and stainless steel vessels should be used, as they are much cheaper and more durable than glassware. The water source and glassware storage area should be in or near the medium preparation area. Work bench tops, suitable for comfortable working while standing should be 85 to 90 cm high and 60 cm deep. The workbench tops should be made with plastic laminate surfaces that can tolerate frequent cleaning.

Sterilization Room

The sterilizing room should be in continuation with the media preparation room. The layout must be planned in such a way that it ensures the smooth movement of the containers from the washing to the media preparation and sterilization room. The sterilization room must have walls and floors that can withstand moisture, heat and steam. An exhaust should be fitted to remove the warm and moist air. The exhaust fan should have an outer cover to prevent entry of outside air. The fan cover should open only when the fan is in operation. In small tissue culture facilities, costly autoclaves can be replaced by simple pressure cookers. However, for large volume media making, horizontal or vertical autoclaves should be installed. Double door autoclaves, which open directly into the media storage room, may be costly but reduce contamination. A cheaper alternative is to transfer the sterilised media to the adjoining room through a hatch window.

Transfer Room

The most important work area is the culture transfer room where the core activity takes place. The transfer area needs to be as clean as possible and be a separate room with minimal air disturbance. Walls and floors of the transfer room must be smooth to ensure frequent cleaning. The doors and windows should be minimal to prevent contamination, but within local safety code. There is no special lighting requirement in the transfer room. The illumination of the laminar airflow chamber is sufficient for work. Sterilization of the instruments can be done with glass-bead sterilizers or flaming after dipping in alcohol, usually ethanol. The culture containers should be stacked on mobile carts (trolleys) to facilitate easy movement from the medium storage room to the transfer room, and finally to the growth room. The chair seats of the transfer operators should be comfortable, as they have to work for long periods in the same place. Fire extinguishers and first aid kits should be provided in the transfer room as a safety measure.

The personnel should leave shoes outside the room. Special laboratory shoes and coats should be worn in this area. Ultraviolet (UV) lights are sometimes installed in transfer areas to disinfect the room; these lights should be used only when people and plant material are not in the room. Safety switches can be installed to turn off the UV lights when regular room lights are turned on.

Growth Room

Growth room is an equally important area where plant cultures are maintained under controlled environmental conditions to achieve optimal growth. It is advisable to have more than one growth room to provide varied culture conditions since different plant species may have different requirements of light and temperature during in vitro culture. Also, in the event of the failure of cooling or lighting in one room, the plant cultures can be moved to another room to prevent loss of cultures.

In the growth room, the number of doors should be minimal to prevent contamination. There is no need for windows in the growth room, except when natural light is used. When artificial lighting is used, the external light can interfere with the photoperiod and temperature of the growth room. Depending on the amount of available space and cost, the culture containers can be placed on either fixed or mobile shelves. Mobile shelves have the advantage of providing access to cultures from both sides of the shelves. The height of the shelves should not exceed 2m. High shelving requires step-up stools to place and remove cultures, being dangerous and time consuming.

The primary source of illumination in the growth room is normally from the lights mounted on the shelves. Overhead light sources can be minimised, as they would be in use only while working during the dark cycle. Plant cultures may not receive uniform light from the conventional downward illumination. Lights directly fitted to the racks create uneven heat distribution. This leads to high humidity within the culture containers, which in turn can cause hyperhydricity. Sideways illumination is an alternative, which requires less number of lights, and provides more uniform lighting. But care has to be taken not to break the lights while moving the cultures across the shelves.

2

Tools and Techniques in Cell Biology

Hands on Training

Biology is a hands on science, and biology students are usually required to spend some time in the laboratory. The type of activities a student will perform vary depending on the exact field they are in. Plant geneticist, for example, sometimes spend time out in fields gathering plants, while molecular biologists may use complex equipment such as DNA sequencing machines. Even so, there are some basic techniques that all beginning lab students should understand.

Basic Skills

- Laboratory safety and understanding warning signs
- Preparation of simple solutions
- Sampling and preparation of samples
- Dissection of plants and animals
- How to clean glassware
- Use of the dissecting scope
- Use of the light microscope
- Making accurate drawings of organisms.

Research Labs

Research laboratories often require more advanced skills such as the following:

- Agarose gel Electrophoresis
- Use of a pH meter
- Use of a spectrophotometer
- Knowledge of distillation techniques
- Colony counting on agar plates
- Cell counting on slides containing grids
- Making a blood smear.

Advanced

Specialized work places will require other skills. Some skills suggested by subject.

Anatomy, Botany, General Biology

- Use of a microtome to make slides
- staining techniques.

Medical

- Cell and tissue culture
- Patient sample handling techniques
- Making an accurate white blood cell count
- Antibody testing methods such as ELISA.

Microbiology

- Use of micropipetter
- Knowledge of Southern and related blotting techniques
- Molecular cloning
- Use of restriction enzymes
- Use of PCR (Polymerase Chain Reaction) machines
- Replica plating
- Growth of competent cells.

Although books exist to teach these techniques, experienced lab workers have special tricks and techniques that they do not write down. That is why it is best to learn by studying with an experienced lab worker.

Mitochondria

In cell biology, a mitochondrion (plural mitochondria) is a membrane-enclosed organelle found in most eukaryotic cells. These organelles range from

0.5 to 10 micrometers (ìm) in diameter. Mitochondria are sometimes described as "cellular power plants" because they generate most of the cell's supply of adenosine triphosphate (ATP), used as a source of chemical energy. In addition to supplying cellular energy, mitochondria are involved in a range of other processes, such as signalling, cellular differentiation, cell death, as well as the control of the cell cycle and cell growth. Mitochondria have been implicated in several human diseases, including mitochondrial disorders and cardiac dysfunction, and may play a role in the aging process.

Several characteristics make mitochondria unique. The number of mitochondria in a cell varies widely by organism and tissue type. Many cells have only a single mitochondrion, whereas others can contain several thousand mitochondria. The organelle is composed of compartments that carry out specialized functions. These compartments or regions include the outer membrane, the intermembrane space, the inner membrane, and the cristae and matrix. Mitochondrial proteins vary depending on the tissue and the species. In humans, 615 distinct types of proteins have been identified from cardiac mitochondria; whereas in Murinae (rats), 940 proteins encoded by distinct genes have been reported. The mitochondrial proteome is thought to be dynamically regulated. Although most of a cell's DNA is contained in the cell nucleus, the mitochondrion has its own independent genome. Further, its DNA shows substantial similarity to bacterial genomes.

Structure

A mitochondrion contains outer and inner membranes composed of phospholipid bilayers and proteins. The two membranes, however, have different properties. Because of this double-membraned organization, there are five distinct compartments within the mitochondrion. There is the outer mitochondrial membrane, the intermembrane space (the space between the outer and inner membranes), the inner mitochondrial membrane, the crista space (formed by infoldings of the inner membrane), and the matrix (space within the inner membrane).

Outer Membrane

The outer mitochondrial membrane, which encloses the entire organelle, has a protein-to-phospholipid ratio similar to that of the eukaryotic plasma membrane (about 1:1 by weight). It contains large numbers of integral proteins called *porins*. These porins form channels that allow molecules 5000 Daltons or less in molecular weight to freely diffuse from one side of the membrane to the other. Larger proteins can enter the mitochondrion if a signalling sequence at their N-terminus binds to a large multisubunit protein called translocase of the outer membrane, which then actively moves them across the membrane. Disruption of the outer membrane permits proteins in the intermembrane space to leak into the cytosol, leading to certain cell death. The mitochondrial outer membrane can associate with the endoplasmic reticulum (ER) membrane, in a structure called MAM (mitochondria-associated ER-membrane). This is important in ER-mitochondria calcium signalling and involved in the transfer of lipids between the ER and mitochondria.

Intermembrane Space

The intermembrane space is the space between the outer membrane and the inner membrane. Because the outer membrane is freely permeable to small molecules, the concentrations of small molecules such as ions and sugars in the intermembrane space is the same as the cytosol. However, large proteins must have a specific signalling sequence to be transported across the outer membrane, the protein composition of this space is different from the protein composition of the cytosol. One protein that is localized to the intermembrane space in this way is cytochrome c.

Inner Membrane

The inner mitochondrial membrane contains proteins with five types of functions:

1. Those that perform the redox reactions of oxidative phosphorylation
2. ATP synthase, which generates ATP in the matrix
3. Specific transport proteins that regulate metabolite passage into and out of the matrix
4. Protein import machinery.
5. Mitochondria fusion and fission protein.

It contains more than 150 different polypeptides, and has a very high protein-to-phospholipid ratio (more than 3:1 by weight, which is about 1 protein for 15 phospholipids). The inner membrane is home to around 1/5 of the total protein in a mitochondrion. In addition, the inner membrane is rich in an unusual phospholipid, cardiolipin. This phospholipid was originally discovered in beef hearts in 1942, and is usually characteristic of mitochondrial and bacterial plasma membranes. Cardiolipin contains four fatty acids rather than two and may help to make the inner membrane impermeable. Unlike the outer membrane, the inner membrane does not contain porins and is highly impermeable to all molecules. Almost all ions and molecules require special membrane transporters to enter or exit the matrix. Proteins are ferried into the matrix via the translocase of the inner membrane (TIM) complex or via Oxa1. In addition, there is a membrane potential across the inner membrane formed by the action of the enzymes of the electron transport chain.

Cristae

The inner mitochondrial membrane is compartmentalized into numerous cristae, which expand the surface area of the inner mitochondrial membrane, enhancing its ability to produce ATP. For typical liver mitochondria the area of the inner membrane is about five times greater than the outer membrane. This ratio is variable and mitochondria from cells that have a greater demand for ATP, such as muscle cells, contain even more cristae. These folds are studded with small round bodies known as F_1 particles or oxysomes. These are not simple random folds but rather invaginations of the inner membrane, which can affect overall chemiosmotic function.

The Matrix

The matrix is the space enclosed by the inner membrane. It contains about two-third of the total protein in a mitochondrion. The matrix is important in the production of ATP with the aid of the ATP synthase contained in the inner membrane. The matrix contains a highly-concentrated mixture of hundreds of enzymes, special mitochondrial ribosomes, tRNA, and several copies of the mitochondrial DNA genome. Of the enzymes, the major functions include oxidation of pyruvate and fatty acids, and the citric acid cycle.

Mitochondria have their own genetic material, and the machinery to manufacture their own RNAs and proteins. A published human mitochondrial DNA sequence revealed 16,569 base pairs encoding 37 total genes: 22 tRNA, 2 rRNA, and 13 peptide genes. The 13 mitochondrial peptides in humans are integrated into the inner mitochondrial membrane, along with proteins encoded by genes that reside in the host cell's nucleus.

Organization and Distribution

Mitochondria are found in nearly all eukaryotes. They vary in number and location according to cell type. A single mitochondrion is often found in unicellular organisms. Conversely, numerous mitochondria are found in human liver cells, with about 1000–2000 mitochondria per cell making up one-fifth of the cell volume. The mitochondria can be found nestled between myofibrils of muscle or wrapped around the sperm flagellum. Often they form a complex 3D branching network inside the cell with the cytoskeleton. The association with the cytoskeleton determines mitochondrial shape, which can affect the function as well. Recent evidence suggests vimentin, one of the components of the cytoskeleton, is critical to the association with the cytoskeleton.

Function

The most prominent roles of mitochondria are to produce ATP (i.e., phosphorylation of ADP) through respiration, and to regulate cellular metabolism. The central set of reactions involved in ATP production are collectively known as the citric acid cycle, or the Krebs Cycle. However, the mitochondrion has many other functions in addition to the production of ATP.

Energy Conversion

A dominant role for the mitochondria is the production of ATP, as reflected by the large number of proteins in the inner membrane for this task. This is done by oxidizing the major products of glucose, pyruvate, and NADH, which are produced in the cytosol.

This process of cellular respiration, also known as aerobic respiration, is dependent on the presence of oxygen. When oxygen is limited, the glycolytic products will be metabolized by anaerobic respiration, a process that is independent of the mitochondria. The production of ATP from glucose has an approximately 13-fold higher yield during aerobic respiration compared to anaerobic respiration. Recently it has been shown that plant mitochondria can produce a limited amount of ATP without oxygen by using the alternate substrate nitrite.

Pyruvate: The Citric Acid Cycle

Each pyruvate molecule produced by glycolysis is actively transported across the inner mitochondrial membrane, and into the matrix where it is oxidized and combined with coenzyme A to form CO_2, acetyl-CoA, and NADH.

The acetyl-CoA is the primary substrate to enter the *citric acid cycle*, also known as the *tricarboxylic acid (TCA) cycle* or *Krebs cycle*. The enzymes of the citric acid cycle are located in the mitochondrial matrix, with the exception of succinate dehydrogenase, which is bound to the inner mitochondrial membrane as part of Complex II. The citric acid cycle oxidizes the acetyl-CoA to carbon dioxide, and, in the process, produces reduced cofactors (three molecules of NADH and one molecule of $FADH_2$) that are a source of electrons for the *electron transport chain*, and a molecule of GTP (that is readily converted to an ATP).

NADH and $FADH_2$: The Electron Transport Chain

The redox energy from NADH and $FADH_2$ is transferred to oxygen (O_2) in several steps via the electron transport chain. These energy-rich molecules are produced within the matrix via the citric acid cycle but are also produced in the cytoplasm by glycolysis. Reducing equivalents from the cytoplasm can be imported via the malate-aspartate shuttle system of antiporter proteins or feed into the electron transport chain using a glycerol phosphate shuttle. Protein complexes in the inner membrane (NADH dehydrogenase, cytochrome c reductase, and cytochrome c oxidase) perform the transfer and the incremental release of energy is used to pump protons (H^+) into the intermembrane space.

This process is efficient, but a small percentage of electrons may prematurely reduce oxygen, forming reactive oxygen species such as superoxide. This can cause oxidative stress in the mitochondria and may contribute to the decline in mitochondrial function associated with the aging process.

As the proton concentration increases in the intermembrane space, a strong electrochemical gradient is established across the inner membrane. The protons can return to the matrix through the ATP synthase complex, and their potential energy is used to synthesize ATP from ADP and inorganic phosphate (P_i). This process is called chemiosmosis, and was first described by Peter Mitchell who was awarded the 1978 Nobel Prize in Chemistry for his work. Later, part of the 1997 Nobel Prize in Chemistry was awarded to Paul D. Boyer and John E. Walker for their clarification of the working mechanism of ATP synthase.

Heat Production

Under certain conditions, protons can re-enter the mitochondrial matrix without contributing to ATP synthesis. This process is known as *proton leak* or *mitochondrial uncoupling* and is due to the facilitated diffusion of protons into the matrix. The process results in the unharnessed potential energy of the proton electrochemical gradient being released as heat. The process is mediated by a proton channel called thermogenin, or UCP1. Thermogenin is a 33kDa protein first discovered in 1973. Thermogenin is primarily found in brown adipose tissue, or brown fat, and is responsible for non-shivering thermogenesis. Brown adipose tissue is found in mammals, and is at its highest levels in early life and in hibernating animals. In humans, brown adipose tissue is present at birth and decreases with age.

Storage of Calcium Ions

The concentrations of free calcium in the cell can regulate an array of reactions and is important for signal transduction in the cell. Mitochondria can transiently store calcium, a contributing process for the cell's homeostasis of calcium. In fact, their ability to rapidly take in calcium for later release makes them very good "cytosolic buffers" for calcium. The endoplasmic reticulum (ER) is the most significant storage site of calcium, and there is a significant interplay between the mitochondrion and ER with regard to calcium. The calcium is taken up into the matrix by a calcium uniporter on the inner mitochondrial membrane. It is primarily driven by the mitochondrial membrane potential. Release of this calcium back into the cell's interior can occur via a sodium-calcium exchange protein or via "calcium-induced-calcium-release" pathways. This can initiate calcium spikes or calcium waves with large changes in the membrane potential. These can activate a series of second messenger system proteins that can coordinate processes such as neurotransmitter release in nerve cells and release of hormones in endocrine cells.

Additional Functions

Mitochondria play a central role in many other metabolic tasks, such as:

- Regulation of the membrane potential
- Apoptosis-programmed cell death
- Calcium signalling (including calcium-evoked apoptosis)
- Cellular proliferation regulation
- Regulation of cellular metabolism
- Certain heme synthesis reactions.
- Steroid synthesis.

Some mitochondrial functions are performed only in specific types of cells. For example, mitochondria in liver cells contain enzymes that allow them to detoxify ammonia, a waste product of protein metabolism. A mutation in the genes regulating any of these functions can result in mitochondrial diseases.

Origin

Mitochondria have many features in common with prokaryotes. As a result, they are believed to be originally derived from endosymbiotic prokaryotes. A mitochondrion contains DNA, which is organized as several copies of a single, circular chromosome. This mitochondrial chromosome contains genes for redox proteins such as those of the respiratory chain. The CoRR Hypothesis proposes that this Co-location is required for Redox Regulation.

The mitochondrial genome codes for some RNAs of ribosomes, and the twenty-two tRNAs necessary for the translation of messenger RNAs into protein. The circular structure is also found in prokaryotes, and the similarity is extended by the fact that mitochondrial DNA is organized with a variant genetic code similar to that of Proteobacteria. This suggests that their ancestor, the so-called proto-mitochondrion, was a member of the Proteobacteria. In particular, the proto-mitochondrion was probably related to the rickettsia. However, the exact relationship of the ancestor of mitochondria to the alpha-proteobacteria and whether the mitochondria was formed at the same time or after the nucleus, remains controversial.

The ribosomes coded for by the mitochondrial DNA are similar to those from bacteria in size and structure. They closely resemble the bacterial 70S ribosome and not the 80S cytoplasmic ribosomes, which are coded for by nuclear DNA.

The endosymbiotic relationship of mitochondria with their host cells was popularized by Lynn Margulis. The endosymbiotic hypothesis suggests that mitochondria descended from bacteria that somehow survived endocytosis by another cell, and became incorporated into the cytoplasm. The ability of these bacteria to conduct respiration in host cells that had relied on glycolysis and fermentation would have provided a considerable evolutionary advantage. In a similar manner, host cells with symbiotic bacteria capable of photosynthesis would have had an advantage. The incorporation of symbiotes would have increased the number of environments in which the cells could survive. This symbiotic relationship probably developed 1.7-2 billion years ago.

A few groups of unicellular eukaryotes lack mitochondria: the microsporidians, metamonads, and archamoebae. These groups appear as the most primitive eukaryotes on phylogenetic trees constructed using rRNA information, which once suggested that they appeared before the origin of mitochondria. However, this is now known to be an artifact of long-branch attraction—they are derived groups and retain genes or organelles derived from mitochondria (e.g., mitosomes and hydrogenosomes).

Genome

The human mitochondrial genome is a circular DNA molecule of about 16 kilobases. It encodes 37 genes: 13 for subunits of respiratory complexes I, III, IV and V, 22 for mitochondrial tRNA (for the 20 standard amino acids, plus an extra gene for leucine and serine), and 2 for rRNA. One mitochondrion can contain two to ten copies of its DNA.

As in prokaryotes, there is a very high proportion of coding DNA and an absence of repeats. Mitochondrial genes are transcribed as multigenic transcripts, which are cleaved and polyadenylated to yield mature mRNAs. Not all proteins necessary for mitochondrial function are encoded by the mitochondrial genome; most are coded by genes in the cell nucleus and the corresponding proteins are imported into the mitochondrion. The exact number of genes encoded by the nucleus and the mitochondrial genome differs between species. In general, mitochondrial genomes are circular, although exceptions have been reported. In general, mitochondrial DNA lacks introns, as is the case in the human mitochondrial genome; however, introns have been observed in some eukaryotic mitochondrial DNA, such as that of yeast and protists, including *Dictyostelium discoideum*.

In animals the mitochondrial genome is typically a single circular chromosome that is approximately 16-kb long and has 37 genes. The genes while highly conserved may vary in location. Curiously this pattern is not found in the human body louse (*Pediculus humanus*). Instead this mitochondrial genome is arranged in 18 minicircular chromosomes each of which is 3–4 kb long and has one to three genes. This pattern is also found in other sucking lice but not in chewing lice. Recombination has been shown to occur between the minichromosomes. The reason for this difference is not known.

While slight variations on the standard code had been predicted earlier, none was discovered until 1979, when researchers studying human mitochondrial genes determined that they used an alternative code. Many slight variants have been discovered since, including various alternative mitochondrial codes. Further, the AUA, AUC, and AUU codons are all allowable start codons.

Exceptions to the universal genetic code (UGC) in mitochondria

Organism	*Codon*	*Standard*	*Novel*
Mammalian	AGA, AGG	Arginine	Stop codon
	AUA	Isoleucine	Methionine
	UGA	Stop codon	Tryptophan
Invertebrates	AGA, AGG	Arginine	Serine
	AUA	Isoleucine	Methionine
	UGA	Stop codon	Tryptophan
Yeast	AUA	Isoleucine	Methionine
	UGA	Stop codon	Tryptophan
	CUA	Leucine	Threonine

Some of these differences should be regarded as pseudo-changes in the genetic code due to the phenomenon of RNA editing, which is common in mitochondria. In higher plants, it was thought that CGG encoded for tryptophan and not arginine; however, the codon in the processed RNA was discovered to be the UGG codon, consistent with the universal genetic code for tryptophan. Of note, the arthropod mitochondrial genetic code has undergone parallel evolution within a phylum, with some organisms uniquely translating AGG to lysine.

Mitochondrial genomes have far fewer genes than the bacteria from which they are thought to be descended. Although some have been lost altogether, many have been transferred to the nucleus, such as the respiratory complex II protein subunits. This is thought to be relatively common over evolutionary time. A few organisms, such as the *Cryptosporidium*, actually have mitochondria that lack any DNA, presumably because all their genes have been lost or transferred. In *Cryptosporidium*, the mitochondria have an altered ATP generation system that renders the parasite resistant to many classical mitochondrial inhibitors such as cyanide, azide, and atovaquone.

Replication and Inheritance

Mitochondria divide by binary fission similar to bacterial cell division; unlike bacteria, however, mitochondria can also fuse with other mitochondria.. The regulation of this division differs between eukaryotes. In many single-celled eukaryotes, their growth and division is linked to the cell cycle. For example, a single mitochondrion may divide synchronously with the nucleus. This division and segregation process must be tightly controlled so that each daughter cell receives at least one mitochondrion. In other eukaryotes (in mammals for example), mitochondria may replicate their DNA and divide mainly in response to the energy needs of the cell, rather than in phase with the cell cycle. When the energy needs of a cell are high, mitochondria grow and divide. When the energy use is low, mitochondria are destroyed or become inactive. In such examples, and in contrast to the situation in many single celled eukaryotes, mitochondria are apparently randomly distributed to the daughter cells during the division of the cytoplasm.

An individual's mitochondrial genes are not inherited by the same mechanism as nuclear genes. At fertilization of an egg cell by a sperm, the egg nucleus and sperm nucleus each contribute equally to the genetic make-up of the zygote nucleus. In contrast, the mitochondria, and therefore the mitochondrial DNA, usually comes from the egg only. The sperm's mitochondria enter the egg but do not contribute genetic information to the embryo. Instead, paternal mitochondria are marked with ubiquitin to select them for later destruction inside the embryo. The egg cell contains relatively few mitochondria, but it is these mitochondria that survive and divide to populate the cells of the adult organism. Mitochondria are, therefore, in most cases inherited down the female line, known as maternal inheritance. This mode is seen in most organisms including all animals. However, mitochondria in some species can sometimes be inherited paternally. This is the norm among certain coniferous plants, although not in pine trees and yew trees. It has been suggested that it occurs at a very low level in humans.

Uniparental inheritance leads to little opportunity for genetic recombination between different lineages of mitochondria, although a single mitochondrion can contain 2–10 copies of its DNA. For this reason, mitochondrial DNA usually is thought to reproduce by binary fission. What recombination does take place maintains genetic integrity rather than maintaining diversity. However, there are studies showing evidence of recombination in mitochondrial DNA. It is clear that the enzymes necessary for recombination are present in mammalian cells. Further, evidence suggests that animal mitochondria can undergo recombination. The data are a bit more controversial in humans, although indirect evidence of recombination exists. If recombination does not occur, the whole mitochondrial DNA sequence represents a single haplotype, which makes it useful for studying the evolutionary history of populations.

Population Genetic Studies

The near-absence of genetic recombination in mitochondrial DNA makes it a useful source of information for scientists involved in population genetics and evolutionary biology. Because all the mitochondrial DNA is inherited as a single unit, or haplotype, the relationships between mitochondrial DNA from different individuals can be represented as a gene tree. Patterns in these gene trees can be used to infer the evolutionary history of populations. The classic example of this is in human evolutionary genetics, where the molecular clock can be used to provide a recent date for mitochondrial Eve. This is often interpreted as strong support for a recent modern human expansion out of Africa. Another human example is the sequencing of mitochondrial DNA from Neanderthal bones. The relatively large evolutionary distance between the mitochondrial DNA sequences of Neanderthals and living humans has been interpreted as evidence for lack of interbreeding between Neanderthals and anatomically-modern humans.

However, mitochondrial DNA reflects the history of only females in a population and so may not represent the history of the population as a whole. This can be partially overcome by the use of paternal genetic sequences, such as the non-recombining region of the Y-chromosome. In a broader sense, only studies that also include nuclear DNA can provide a comprehensive evolutionary history of a population.

Dysfunction and Disease

Mitochondrial Diseases

With their central place in cell metabolism, damage — and subsequent dysfunction — in mitochondria is an important factor in a wide range of human diseases. Mitochondrial disorders often present as neurological disorders, but can manifest as myopathy, diabetes, multiple endocrinopathy, or a variety of other systemic manifestations. Diseases caused by mutation in the mtDNA include Kearns-Sayre syndrome, MELAS syndrome and Leber's hereditary optic neuropathy. In the vast majority of cases, these diseases are transmitted by a female to her children, as the zygote derives its mitochondria and hence its mtDNA from the ovum. Diseases such as Kearns-Sayre syndrome, Pearson's syndrome, and progressive external ophthalmoplegia are thought to be due to large-scale mtDNA rearrangements, whereas other diseases such as MELAS syndrome, Leber's hereditary optic neuropathy, myoclonic epilepsy with ragged red fibres (MERRF), and others are due to point mutations in mtDNA. In other diseases, defects in nuclear genes lead to dysfunction of mitochondrial proteins. This is the case in Friedreich's ataxia, hereditary spastic paraplegia, and Wilson's disease. These diseases are inherited in a dominance relationship, as applies to most other genetic diseases. A variety of disorders can be caused by nuclear mutations of oxidative phosphorylation enzymes, such as coenzyme Q10 deficiency and Barth syndrome. Environmental influences may interact with hereditary predispositions and cause mitochondrial disease. For example, there may be a link between pesticide exposure and the later onset of Parkinson's disease.

Other pathologies with etiology involving mitochondrial dysfunction include schizophrenia, bipolar disorder, dementia, Alzheimer's disease, Parkinson's disease, epilepsy, stroke, cardiovascular disease, retinitis pigmentosa, and diabetes mellitus. A common thread thought to link these seemingly-unrelated conditions is cellular damage causing oxidative stress. How exactly mitochondrial dysfunction fits into the etiology of these pathologies is yet to be elucidated.

Possible Relationships to Aging

Given the role of mitochondria as the cell's powerhouse, there may be some leakage of the high-energy electrons in the respiratory chain to form reactive oxygen species. This can result in significant oxidative stress in the mitochondria with high mutation rates of mitochondrial DNA. A vicious cycle is thought to occur, as oxidative stress leads to mitochondrial DNA mutations, which can lead to enzymatic abnormalities and further oxidative stress. A number of changes occur to mitochondria during the aging process. Tissues from elderly patients show a decrease in enzymatic activity of the proteins of the respiratory chain. Large deletions in the mitochondrial genome can lead to high levels of oxidative stress and neuronal death in Parkinson's disease. Hypothesized links between aging and oxidative stress are not new and were proposed over 50 years ago; however, there is much debate over whether mitochondrial changes are causes of aging or merely characteristics of aging. One notable study in mice demonstrated shortened lifespan but no increase in reactive oxygen species despite increasing mitochondrial DNA mutations, suggesting that mitochondrial DNA mutations can cause lifespan shortening by other mechanisms. As a result, the exact relationships between mitochondria, oxidative stress, and aging have not yet been settled.

Plastids

Plastids are major organelles found in the cells of plants and algae. Plastids are the site of manufacture and storage of important chemical compounds used by the cell. Plastids often contain pigments used in photosynthesis, and the types of pigments present can change or determine the cell's color.

Plastids in Plants

Plastids are responsible for photosynthesis, storage of products like starch and for the synthesis have the ability to differentiate, or redifferentiate, between these and other forms. All plastids are derived from proplastids (formerly "eoplasts", *eo*-: dawn, early), which are present in the meristematic regions of the plant. Proplastids and young chloroplasts commonly divide, but more mature chloroplasts also have this capacity. In plants, plastids may differentiate into several forms, depending upon which function they need to play in the cell. Undifferentiated plastids (*proplastids*) may develop into any of the following plastids:

- Chloroplasts: for photosynthesis; *see also etioplasts, the predecessors of chloroplasts*
- Chromoplasts: for pigment synthesis and storage
- Gerontoplasts: control the dismantling of the photosynthetic apparatus during senescence
- Leucoplasts: for monoterpene synthesis; *leucoplasts sometimes differentiate into more specialized plastids:*
    - Amyloplasts: for starch storage and detecting gravity
    - Elaioplasts: for storing fat
    - Proteinoplasts: for storing and modifying protein.

Each plastid creates multiple copies of the circular 75-250 kilobase plastome. The number of genome copies per plastid is flexible, ranging from more than 1000 in rapidly dividing cells, which generally contain few plastids, to 100 or fewer in mature cells, where plastid divisions has given rise to a large number of plastids. The plastome contains about 100 genes encoding ribosomal and transfer ribonucleic acids (rRNAs and tRNAs) as well as proteins involved in photosynthesis and plastid gene transcription and translation. However, these proteins only represent a small fraction of the total protein set-up necessary to build and maintain the structure and function of a particular type of plastid. Nuclear genes encode the vast majority of plastid proteins, and the expression of plastid genes and nuclear genes is tightly co-regulated to allow proper development of plastids in relation to cell differentiation.

Plastid DNA exists as large protein-DNA complexes associated with the inner envelope membrane and called 'plastid nucleoids'. Each nucleoid particle may contain more than 10 copies of the plastid DNA. The proplastid contains a single nucleoid located in the centre of the plastid. The developing plastid has many nucleoids, localized at the periphery of the plastid, bound to the inner envelope membrane. During the development of proplastids to chloroplasts, and when plastids convert from one type to another, nucleoids change in morphology, size and location within the organelle. The remodelling of nucleoids is believed to occur by modifications to the composition and abundance of nucleoid proteins. Many plastids, particularly those responsible for photosynthesis, possess numerous internal membrane layers. In plant cells, long thin protuberances called stromules sometimes form and extend from the main plastid body into the cytosol and interconnect several plastids. Proteins, and presumably smaller molecules, can move within stromules. Most cultured cells that are relatively large compared to other plant cells have very long and abundant stromules that extend to the cell periphery.

Plastids in Algae

In algae, the term leucoplast is used for all unpigmented plastids and their function differs from the leucoplasts of plants. Etioplasts, amyloplasts and chromoplasts are plant-specific and do not occur in algae. Plastids in algae and hornworts may also differ from plant plastids in that they contain pyrenoids.

Glaucocystophytic algae contain muroplasts which are similar to chloroplasts except that they have a cell wall that is similar to that of prokaryotes. Rhydophytic algae contain rhydoplasts which are red chloroplasts which allow the algae to photosynthesise to a depth of up to 268m.

Inheritance of Plastids

Most plants inherit the plastids from only one parent. Angiosperms generally inherit plastids from the female gamete, while many gymnosperms inherit plastids from the male pollen. Algae also inherit plastids from only one parent. The plastid DNA of the other parent is thus completely lost.

In normal intraspecific crossings (resulting in normal hybrids of one species), the inheritance of plastid DNA appears to be quite strictly 100% uniparental. In interspecific hybridisations, however, the inheritance of plastids appears to be more erratic. Although plastids inherit mainly maternally in interspecific hybridisations, there are many reports of hybrids of flowering plants that contain plastids of the father. Approximately ~20% of angiosperms, including alfalfa (Medicago), normally show biparental inheritance of plastids.

Origin of Plastids

Plastids are thought to have originated from endosymbiotic cyanobacteria. They developed around 1500 million years ago and allowed eukaryotes to carry out oxygenic photosynthesis. Due to a split-up into three evolutionary lineages, the plastids are named differently: chloroplasts in green algae and plants, rhodoplasts in red algae and cyanelles in the glaucophytes. The plastids differ by their pigmentation, but also in ultrastructure. The chloroplasts e.g. have lost all phycobilisomes, the light harvesting complexes found in cyanobacteria, red algae and glaucophytes, but only in plants and in closely related green algae contain stroma and grana thylakoids. The glaucocystophycean plastid in contrast to the chloroplasts and the rhodoplasts is still surrounded by the remains of the cyanobacterial cell wall. All these primary plastids are surrounded by two membranes.

Complex plastids start by secondary endosymbiosis, when a eukaryote engulfs a red or green alga and retains the algal plastid, which is typically surrounded by more than two membranes. In some cases these plastids may be reduced in their metabolic and/or photosynthetic capacity. Algae with complex plastids derived by secondary endosymbiosis of a red alga include the heterokonts, haptophytes, cryptomonads, and most dinoflagellates (= rhodoplasts). Those that endosymbiosed a green alga include the euglenids and chlorarachniophytes (= chloroplasts). The Apicomplexa, a phylum of obligate parasitic protozoa including the causative agents of malaria (*Plasmodium* spp.), toxoplasmosis (*Toxoplasma gondii*), and many other human or animal diseases also harbour a complex plastid (although this organelle has been lost in some apicomplexans, such as *Cryptosporidium parvum*, which causes cryptosporidiosis). The 'apicoplast' is no longer capable of photosynthesis, but is' an essential organelle, and a promising target for antiparasitic drug development. Some dinoflagellates and sea slugs, particularly of the genus Elysia, take up algae as food and keep the plastid of the digested alga to profit from the photosynthesis; after a while the plastids are also digested. These captured plastids are known as kleptoplastids.

Endoplasmic Reticulum

The endoplasmic reticulum (ER) is an eukaryotic organelle that forms an interconnected network of tubules, vesicles, and cisternae within cells. Rough endoplasmic reticulums synthesize proteins, while smooth endoplasmic reticulums synthesize lipids and steroids, metabolize carbohydrates and steroids, and regulate calcium concentration, drug detoxification, and attachment of receptors on cell membrane proteins. Sarcoplasmic reticulums solely regulate calcium levels. The lacey membranes of the endoplasmic reticulum were first seen by Keith R. Porter, Albert Claude, and Ernest F. Fullam in 1945.

Structure

The general structure of the endoplasmic reticulum is an extensive membrane network of cisternae (sac-like structures) held together by the cytoskeleton. The phospholipid membrane encloses a space, the cisternal space (or lumen), from the cytosol, which is continuous with the perinuclear space. The functions of the endoplasmic reticulum vary greatly depending on the exact type of endoplasmic reticulum and the type of cell in which it resides. The three varieties are called *rough endoplasmic reticulum*, *smooth endoplasmic reticulum* and *sarcoplasmic reticulum*.

The quantity of RER and SER in a cell can quickly interchange from one type to the other, depending on changing metabolic needs: one type will undergo numerous changes including new proteins embedded in the membranes in order to transform. Also, massive changes in the protein content can occur without any noticeable structural changes, depending on the enzymatic needs of the cell (as per the functions listed below).

Rough Endoplasmic Reticulum

The surface of the rough endoplasmic reticulum (RER) is studded with protein-manufacturing ribosomes giving it a "rough" appearance (hence its name). However, the ribosomes bound to the RER at any one time are not a stable part of this organelle's structure as ribosomes are constantly being bound and released from the membrane. A ribosome only binds to the ER once it begins to synthesize a protein destined for the secretory pathway. Here, a ribosome in the cytosol begins synthesizing a protein until a signal recognition particle recognizes the pre-piece of 5-15 hydrophobic amino acids preceded by a positively charged amino acid. This signal sequence allows the recognition particle to bind to the ribosome, causing the ribosome to bind to the RER and pass the new protein through the ER membrane. The pre-piece is then cleaved off within the lumen of the ER and the ribosome released back into the cytosol.

The membrane of the RER is continuous with the outer layer of the nuclear envelope. Although there is no continuous membrane between the RER and the Golgi apparatus, membrane-bound vesicles shuttle proteins between these two compartments. Vesicles are surrounded by coating proteins called COPI and COPII. COPII targets vesicles to the golgi and COPI marks them to be brought back to the RER. The RER works in concert with the Golgi complex to target new proteins to their proper destinations. A second method of transport out of the ER are areas called membrane contact sites, where the membranes of the ER and other organelles are held closely together, allowing the transfer of lipids and other small molecules.

The RER is key in multiple functions:

- lysosomal enzymes with a mannose-6-phosphate marker added in the *cis*-Golgi network

- Secreted proteins, either secreted constitutively with no tag, or regulated secretion involving clathrin and paired basic amino acids in the signal peptide.
- integral membrane proteins that stay imbedded in the membrane as vesicles exit and bind to new membranes. Rab proteins are key in targeting the membrane, SNAP and SNARE proteins are key in the fusion event.
- initial glycosylation as assembly continues. This is either N-linked (O-linking occur in the golgi).

o N-linked glycosylation: if the protein is properly folded, glycosyl-transferase recognizes the AA sequence NXS or NXT (with the S/T residue phosphorylated) and adds a 14 sugar backbone (2 *N*-acetylglucosamine, 9 branching mannose; and 3 glucose at the end) to the side chain nitrogen of Asn.

Smooth Endoplasmic Reticulum

The smooth endoplasmic reticulum (SER) has functions in several metabolic processes, including synthesis of lipids and steroids, metabolism of carbohydrates, regulation of calcium concentration, drug detoxification, attachment of receptors on cell membrane proteins, and steroid metabolism. It is connected to the nuclear envelope. Smooth endoplasmic reticulum is found in a variety of cell types (both animal and plant) and it serves different functions in each. The Smooth ER also contains the enzyme glucose-6-phosphatase which converts glucose-6-phosphate to glucose, a step in gluconeogenesis. The SER consists of tubules and vesicles that branch forming a network. In some cells there are dilated areas like the sacs of RER. The network of SER allows increased surface area for the action or storage of key enzymes and the products of these enzymes.

Sarcoplasmic Reticulum

The sarcoplasmic reticulum (SR), from the Greek *sarx*, ("flesh"), is a special type of smooth ER found in smooth and striated muscle. The only structural difference between this organelle and the SER is the medley of proteins they have, both bound to their membranes and drifting within the confines of their lumens. This fundamental difference is indicative of their functions: the SER synthesizes molecules while the SR stores and pumps calcium ions. The SR contains large stores of calcium, which it sequesters and then releases when the muscle cell is stimulated. The SR's release of calcium upon electrical stimulation of the cell plays a major role in excitation-contraction coupling.

Functions

The endoplasmic reticulum serves many general functions, including the facilitation of protein folding and the transport of synthesized proteins in sacs called cisternae. Correct folding of newly-made proteins is made possible by several endoplasmic reticulum chaperone proteins, including protein disulfide isomerase (PDI), ERp29, the Hsp70 family member Grp78, calnexin, calreticulin, and the peptidylpropyl isomerase family. Only properly-folded proteins are transported from the rough ER to the Golgi complex.

Transport of Proteins

Secretory proteins, mostly glycoproteins, are moved across the endoplasmic reticulum membrane. Proteins that are transported by the endoplasmic reticulum and from there throughout the cell are marked with an address tag called a signal sequence.

The N-terminus (one end) of a polypeptide chain (i.e., a protein) contains a few amino acids that work as an address tag, which are removed when the polypeptide reaches its destination. Proteins that are destined for places outside the endoplasmic reticulum are packed into transport vesicles and moved along the cytoskeleton toward their destination.

The endoplasmic reticulum is also part of a protein sorting pathway. It is, in essence, the transportation system of the eukaryotic cell. The majority of endoplasmic reticulum resident proteins are retained in the endoplasmic reticulum through a retention motif. This motif is composed of four amino acids at the end of the protein sequence. The most common retention sequence is KDEL (*lys-asp-glu-leu*). However, variation on KDEL does occur and other sequences can also give rise to endoplasmic reticulum retention. It is not known if such variation can lead to sub-endoplasmic reticulum localizations. There are three KDEL receptors in mammalian cells, and they have a very high degree of sequence identity. The functional differences between these receptors remain to be established.

Other Functions

- Insertion of proteins into the endoplasmic reticulum membrane: Integral membrane proteins are inserted into the endoplasmic reticulum membrane as they are being synthesized (co-translational translocation). Insertion into the endoplasmic reticulum membrane requires the correct topogenic signal sequences in the protein.
- Glycosylation: Glycosylation involves the attachment of oligosaccharides.
- Disulfide bond formation and rearrangement: Disulfide bonds stabilize the tertiary and quaternary structure of many proteins.

- Drug Metabolism: The smooth ER is the site at which some drugs are modified by microsomal enzymes which include the cytochrome P450 enzymes.

Golgi Complex

Golgi apparatus (also Golgi body or Golgi Complex) is an organelle found in most eukaryotic cells. It was identified in 1897 by the Italian physician Camillo Golgi and was named after him. The primary function of the Golgi apparatus is to process and package macromolecules, such as proteins and lipids, after their synthesis and before they make their way to their destination; it is particularly important in the processing of proteins for secretion. The Golgi apparatus forms a part of the cellular endomembrane system.

Discovery

Due to its fairly large size, the golgi apparatus was one of the first organelles to be discovered and observed in detail. The apparatus was discovered in 1897 by Italian physician Camillo Golgi during an investigation of the nervous system. After first observing it under his microscope, he termed the structure the *internal reticular apparatus*. The structure was then renamed after Golgi not long after the announcement of his discovery in 1898. However, some doubted the discovery at first, arguing that the appearance of the structure was merely an optical illusion created by the observation technique used by Golgi. With the development of modern microscopes in the 20th century, the discovery was confirmed.

Structure

The Golgi is composed of stacks of membrane-bound structures known as cisternae (singular: *cisterna*). An individual stack is sometimes called a dictyosome (from Greek dictyon, net + soma, body), especially in plant cells. A mammalian cell typically contains 40 to 100 stacks. Between four and eight cisternae are usually present in a stack; however, in some protists as many as sixty have been observed. Each cisterna comprises a flattened membrane disk, and carries Golgi enzymes to help or to modify cargo proteins that travel through them. They are found in both plant and animal cells. The cisternae stack has four functional regions: the cis-Golgi network, medial-Golgi, endo-Golgi, and trans-Golgi network. Vesicles from the endoplasmic reticulum (via the vesicular-tubular clusters) fuse with the network and subsequently progress through the stack to the trans Golgi network, where they are packaged and sent to the required destination. Each region contains different enzymes which selectively modify the contents depending on where they reside. The cisternae also carry structural proteins important for their maintenance as flattened membranes which stack upon each other.

Function

Cells synthesize a large number of different macromolecules. The Golgi apparatus is integral in modifying, sorting, and packaging these macromolecules for cell secretion (exocytosis) or use within the cell. It primarily modifies proteins delivered from the rough endoplasmic reticulum but is also involved in the transport of lipids around the cell, and the creation of lysosomes. In this respect it can be thought of as similar to a post office; it packages and labels items which it then sends to different parts of the cell. Enzymes within the cisternae are able to modify substances by the addition of carbohydrates (glycosylation) and phosphates (phosphorylation). In order to do so, the Golgi imports substances such as nucleotide sugars from the cytosol. These modifications may also form a signal sequence which determines their final destination. For example, the Golgi apparatus adds a mannose-6-phosphate label to proteins destined for lysosomes.

The Golgi plays an important role in the synthesis of proteoglycans, which are molecules present in the extracellular matrix of animals. It is also a major site of carbohydrate synthesis. This includes the productions of glycosaminoglycans (GAGs), long unbranched polysaccharides which the Golgi then attaches to a protein synthesised in the endoplasmic reticulum to form proteoglycans. Enzymes in the Golgi polymerize several of these GAGs via a xylose link onto the core protein. Another task of the Golgi involves the sulfation of certain molecules passing through its lumen via sulphotranferases that gain their sulphur molecule from a donor called PAPs. This process occurs on the GAGs of proteoglycans as well as on the core protein. The level of sulfation is very important to the proteoglycans' signalling abilities as well as giving the proteoglycan its overall negative charge.

The phosphorylation of molecules requires that ATP is imported into the lumen of the Golgi and then utilised by resident kinases such as casein kinase 1 and casein kinase 2. One molecule that is phosphorylated in the Golgi is Apolipoprotein, which forms a molecule known as VLDL that is a constitute of blood serum. It is thought that the phosphorylation of these molecules is important to help aid in their sorting for secretion into the blood serum. The Golgi has a putative role in apoptosis, with several Bcl-2 family members localised there, as well as to the mitochondria. A newly characterized protein, GAAP (Golgi anti-apoptotic protein), almost exclusively resides in the Golgi and protects cells from apoptosis by an as-yet undefined mechanism.

Vesicular Transport

The vesicles that leave the rough endoplasmic reticulum are transported to the *cis* face of the Golgi apparatus, where they fuse with the Golgi membrane and empty their contents into the lumen. Once inside they are modified, sorted and shipped towards their final destination. As such, the Golgi apparatus tends to be more prominent and numerous in cells synthesising and secreting many substances: plasma B cells, the antibody-secreting cells of the immune system, have prominent Golgi complexes. Those proteins destined for areas of the cell other than either the endoplasmic reticulum or Golgi apparatus are moved towards the *trans* face, to a complex network of membranes and associated vesicles known as the *trans-Golgi network* (TGN). This area of the Golgi is the point at which proteins are sorted and shipped to their intended destinations by their placement into one of at least three different types of vesicles, depending upon the molecular marker they carry:

Transport Mechanism

The transport mechanism which proteins use to progress through the Golgi apparatus is not yet clear; however a number of hypotheses currently exist. Until recently, the vesicular transport mechanism was favoured but now more evidence is coming to light to support cisternal maturation. The two proposed models may actually work in conjunction with each other, rather than being mutually exclusive. This is sometimes referred to as the *combined* model.

- *Cisternal maturation model*: the cisternae of the Golgi apparatus move by being built at the *cis* face and destroyed at the *trans* face. Vesicles from the endoplasmic reticulum fuse with each other to form a cisterna at the *cis* face, consequently this cisterna would appear to move through the Golgi stack when a new cisterna is formed at the *cis* face. This model is supported by the fact that structures larger than the transport vesicles, such as collagen rods, were observed microscopically to progress through the Golgi apparatus. This was initially a popular hypothesis, but lost favour in the 1980s. Recently it has made a comeback, as laboratories at the University of Chicago and the University of Tokyo have been able to use new technology to directly observe Golgi compartments maturing. Additional evidence comes from the fact that COPI vesicles move in the retrograde direction, transporting Endoplasmic Reticulum proteins back to where they belong by recognizing a signal peptide.
- *Vesicular transport model*: Vesicular transport views the Golgi as a very stable organelle, divided into compartments in the cis to trans direction. Membrane bound carriers transport material between the ER and the different compartments of the Golgi. Experimental evidence includes the abundance of small vesicles (known technically as shuttle vesicles) in proximity to the Golgi apparatus. To direct the vesicles,

actin filaments connect packaging proteins to the membrane to ensure that they fuse with the correct compartment.

Golgi Apparatus During Mitosis

The Golgi apparatus will break up and disappear following the onset of mitosis, or cellular division. During the telophase of mitosis, the Golgi apparatus reappears; however, it is still uncertain how this occurs.

Golgi Apparatus in Popular Culture

"Golgi Apparatus" is the title of a song by the band Phish in their album Junta. Tom Marshall and guitarist Trey Anastasio wrote this with some friends during grade school and later reworked it as a Phish song.

Lysosomes

Lysosomes are cube shaped organelles that contain enzymes (acid hydrolases) that break up waste materials and cellular debris. They are found in animal cells, while in yeast and plants the same roles are performed by lytic vacuoles. Lysosomes digest excess or worn-out organelles, food particles, and engulfed viruses or bacteria. The membrane around a lysosome allows the digestive enzymes to work at the 4.5 pH they require. Lysosomes fuse with vacuoles and dispense their enzymes into the vacuoles, digesting their contents. They are created by the addition of hydrolytic enzymes to early endosomes from the Golgi apparatus. The name *lysosome* derives from the Greek words *lysis*, which means to separate; and *soma*, which means body. They are frequently nicknamed "suicide-bags" or "suicide-sacs" by cell biologists due to their role in autolysis. Lysosomes were discovered by the Belgian cytologist Christian de Duve in 1949.

The size of lysosomes varies from 0.1–1.2 m. At pH 4.8, the interior of the lysosomes is acidic compared to the slightly alkaline cytosol (pH 7.2). The lysosome maintains this pH differential by pumping protons (H ions) from the cytosol across the membrane via proton pumps and chloride ion channels. The lysosomal membrane protects the cytosol, and therefore the rest of the cell, from the degradative enzymes within the lysosome. The cell is additionally protected from any lysosomal acid hydrolases that leak into the cytosol as these enzymes are pH-sensitive and function less well in the alkaline environment of the cytosol.

Enzymes

Some important enzymes found within lysosomes include:

- Lipase, which digests lipids
- Amylase, which digest Starchs (a kind of carbohydrate)

- Proteases, which digest proteins
- Nucleases, which digest nucleic acids
- phosphoric acid monoesters.

Lysosomal enzymes are synthesized in the cytosol and the endoplasmic reticulum, where they receive a mannose-6-phosphate tag that targets them for the lysosome. Aberrant lysosomal targeting causes inclusion-cell disease, whereby enzymes do not properly reach the lysosome, resulting in accumulation of waste within these organelles.

Functions

Lysosomes are the cell's garbage disposal system. They are used for the digestion of macromolecules from phagocytosis (ingestion of other dying cells or larger extracellular material, like foreign invading microbes), endocytosis (where receptor proteins are recycled from the cell surface), and autophagy (where in old or unneeded organelles or proteins, or microbes that have invaded the cytoplasm are delivered to the lysosome). Autophagy may also lead to autophagic cell death, a form of programmed self-destruction, or autolysis, of the cell, which means that the cell is digesting itself.

Other functions include digesting foreign bacteria (or other forms of waste) that invade a cell and helping repair damage to the plasma membrane by serving as a membrane patch, sealing the wound. In the past, lysosomes were thought to kill cells that were no longer wanted, such as those in the tails of tadpoles or in the web from the fingers of a 3- to 6-month-old fetus. While lysosomes digest some materials in this process, it is actually accomplished through programmed cell death, called apoptosis.

Clinical Relevance

There are a number of lysosomal storage diseases that are caused by the malfunction of the lysosomes or one of their digestive proteins; examples include Tay-Sachs disease and Pompe's disease. These diseases are caused by a defective or missing digestive protein, which leads to the accumulation of substrates within the cell, impairing metabolism.

In the broad sense, these can be classified as mucopolysaccharidoses, GM_2 gangliosidoses, lipid storage disorders, glycoproteinoses, mucolipidoses, or leukodystrophies.

3

Cell Reproduction: Cell Growth and Cell Cycle

The process of a cell splitting and becoming two similar cells.

Prokaryotes reproduce in a process called binary fission.

Eukaryotic cells reproduce using either mitosis or meiosis. Mitosis creates two daughter cells with an equal number of chromosomes. Meiosis creates four daughter cells, each with half the number chromosomes as the parent, and is used in sexual reproduction.

Cell Division

Cell division is the process by which a *parent cell* divides into two or more *daughter cells*. Cell division is usually a small segment of a larger cell cycle. This type of cell division in eukaryotes is known as mitosis, and leaves the daughter cell capable of dividing again. The corresponding sort of cell division in prokaryotes is known as binary fission. In another type of cell division present only in eukaryotes, called meiosis, a cell is permanently transformed into a gamete and cannot divide again until fertilization. Right before the parent cell splits, it undergoes DNA replication.

For simple unicellular organisms such as the amoeba, one cell division is equivalent to reproduction— an entire new organism is created. On a larger scale, mitotic cell division can create progeny from multicellular organisms, such as plants that grow from cuttings. Cell division also enables asexually reproducing organisms to develop from the one-celled zygote, which itself was produced by cell division from gametes. And after growth, cell division allows for continual construction and repair of the organism.

A human being's body experiences about 10,000 trillion cell divisions in a lifetime. The primary concern of cell division is the maintenance of the original cell's genome.

Before division can occur, the genomic information which is stored in chromosomes must be replicated, and the duplicated genome separated cleanly between cells. A great deal of cellular infrastructure is involved in keeping genomic information consistent between "generations".

Variants

Cells are classified into two categories: simple, non-nucleated prokaryotic cells, and complex, nucleated eukaryotic cells. By dint of their structural differences, eukaryotic and prokaryotic cells do not divide in the same way.

Furthermore, the pattern of cell division that transforms eukaryotic stem cells into gametes (sperm in males or ova in females) is different from that of eukaryotic somatic (non-germ) cells.

Degradation

Multicellular organisms replace worn-out cells through cell division. In some animals, however, cell division eventually halts. In humans this occurs on average, after 52 divisions, known as the Hayflick limit. The cell is then referred to as senescent. Cells stop dividing because the telomeres, protective bits of DNA on the end of a chromosome required for replication, shorten with each copy, eventually being consumed, as described in the article on telomere shortening. Cancer cells, on the other hand, are not thought to degrade in this way, if at all. An enzyme called telomerase, present in large quantities in cancerous cells, rebuilds the telomeres, allowing division to continue indefinitely.

Cell Growth

The term cell growth is used in the contexts of cell development and cell division (reproduction) When used in the context of cell division, it refers to growth of cell populations, where one cell (the "mother cell") grows and divides to produce two "daughter cells".

Cell Populations

Cell populations go through a particular type of exponential growth called doubling. Thus, each generation of cells should be twice as numerous as the previous generation. However, the number of generations only gives a maximum figure as not all cells survive in each generation.

Cell Size

Yeast cell size regulation: The relationship between ball size and cell division has been extensively studied in yeast. For some cells, there is a mechanism by which cell division is not initiated until a cell has reached a certain size. If the nutrient supply is restricted (after time t = 2 in the diagram, below), and the rate of increase in cell size is slowed, the time period between cell divisions is increased. Yeast cell size mutants were isolated that begin cell division before reaching the normal size (*wee* mutants). The Wee1 protein is a tyrosine kinase. It normally phosphorylates the Cdc2 cell cycle regulatory protein (cyclin-dependent kinase-1, CDK1) on a tyrosine residue. This covalent modification of the molecular structure of Cdc2 inhibits the enzymatic activity of Cdc2 and prevents cell division. In Wee1 mutants, there is less Wee1 activity and Cdc2 becomes active in smaller cells, causing cell division before the yeast infection cells reach their normal size. Cell division may be regulated in part by dilution of Wee1 protein in cells as they grow larger.

Cell Size Regulation in Mammals

Many of the signal molecules that convey information to cells during the control of cellular differentiation or growth are called growth factors. The protein mTOR is a serine/threonine kinase that regulates translation and cell division. Nutrient availability influences mTOR so that when cells are not able to grow to normal size they will not undergo cell division. The details of the molecular mechanisms of mammalian cell size control are currently being investigated. The size of post-mitotic neurons depends on the size of the cell body, axon and dendrites. In vertebrates, neuron size is often a reflection of the number of synaptic contacts onto the neuron or from a neuron onto other cells. For example, the size of motoneurons usually reflects the size of the motor unit that is controlled by the motoneuron. Invertebrates often have giant neurons and axons that provide special functions such as rapid action potential propagation. Mammals also use this trick for increasing the speed of signals in the nervous system, but they can also use myelin to accomplish this, so most human neurons are relatively small cells.

Other Experimental Systems for the Study of Cell Size Regulation

One common means to produce very large cells is by cell fusion to form syncytia. For example, very long (several inches) skeletal muscle cells are formed by fusion of thousands of myocytes. Genetic studies of the fruit fly *Drosophila* have revealed several genes that are required for the formation of multinucleated muscle cells by fusion of myoblasts. Some of the key proteins are important for cell adhesion between myocytes and some are involved in adhesion-dependent cell-to-cell signal transduction that allows for a cascade of cell fusion events. Oocytes can be unusually large cells in species for which embryonic development takes place away from the mother's body. Their large size can be achieved either by pumping in cytosolic components from adjacent cells through cytoplasmic bridges (*Drosophila*) or by internalization of nutrient storage granules (yolk granules) by endocytosis (frogs).

Increases in the size of plant cells are complicated by the fact that almost all plant cells are inside of a solid cell wall. Under the influence of certain plant hormones the cell wall can be remodeled, allowing for increases in cell size that are important for the growth of some plant tissues.

Most unicellular organisms are microscopic in size, but there are some giant bacteria and protozoa that are visible to the naked eye.

Cell reproduction is asexual. For most of the constituents of the cell, growth is a steady, continuous process, interrupted only briefly at M phase when the nucleus and then the cell divide in two.

The process of cell division, called cell cycle, has four major parts called phases. The first part, called G_1 phase is marked by synthesis of various enzymes that are required for DNA replication. The second part of the cell cycle is the S phase, where DNA replication produces two identical sets of chromosomes. The third part is the G_2 phase. Significant protein synthesis occurs during this phase, mainly involving the production of microtubules, which are required during the process of division, called mitosis. The fourth phase, M phase, consists of nuclear division (karyokinesis) and cytoplasmic division (cytokinesis), accompanied by the formation of a new cell membrane. This is the physical division of "mother" and "daughter" cells. The M phase has been broken down into several distinct phases, sequentially known as prophase, prometaphase, metaphase, anaphase and telophase leading to cytokinesis.

Cell division is more complex in eukaryotes than in other organisms. Prokaryotic cells such as bacterial cells reproduce by binary fission, a process that includes DNA replication, chromosome segregation, and cytokinesis. Eukaryotic cell division either involves mitosis or a more complex process called meiosis. Mitosis and meiosis are sometimes called the two "nuclear division" processes. Binary fission is similar to eukaryotic cell reproduction that involves mitosis. Both lead to the production of two daughter cells with the same number of chromosomes as the parental cell. Meiosis is used for a special cell reproduction process of diploid organisms. It produces four special daughter cells (gametes) which have half the normal cellular amount of DNA. A male and a female gamete can then combine to produce a zygote, a cell which again has the normal amount of chromosomes.

Comparison of the Three Types of Cell Division

The DNA content of a cell is duplicated at the start of the cell reproduction process. Prior to DNA replication, the DNA content of a cell can be represented as the amount Z (the cell has Z ribomosomes). After the DNA replication process, the amount of DNA in the cell is 2Z (multiplication: 2 x Z = 2Z). During Binary fission and mitosis the duplicated DNA content of the reproducing parental cell is separated into two equal halves that are destined to end up in the two daughter cells. The final part of the cell reproduction process is cell division, when daughter cells physically split apart from a parental cell. During meiosis, there are two cell division steps that together produce the four daughter cells.

After the completion of binary fission or cell reproduction involving mitosis, each daughter cell has the same amount of DNA (Z) as what the parental cell had before it replicated its DNA. These two types of cell reproduction produced two daughter cells that have the same number of chromosomes as the parental cell. After meiotic cell reproduction the four daughter cells have half the number of chromosomes that the parental cell originally had.

This is the haploid amount of DNA, often symbolized as N. Meiosis is used by diploid organisms to produce haploid gametes. In a diploid organism such as the human organism, most cells of the body have the diploid amount of DNA, 2N.

Using this notation for counting chromosomes we say that human somatic cells have 46 chromosomes (2N = 46) while human sperm and eggs have 23 chromosomes (N = 23). Humans have 23 distinct types of chromosomes, the 22 autosomes and the special category of sex chromosomes. There are two distinct sex chromosomes, the X chromosome and the Y chromosome. A diploid human cell has 23 chromosomes from that person's father and 23 from the mother. That is, your body has two copies of human chromosome number 2, one from each of your parents.

Immediately after DNA replication a human cell will have 46 "double chromosomes". In each double chromosome there are two copies of that chromosome's DNA molecule. During mitosis the double chromosomes are split to produce 92 "single chromosomes", half of which go into each daughter cell. During meiosis, there are two chromosome separation steps which assure that each of the four daughter cells gets one copy of each of the 23 types of chromosome.

Sexual Reproduction

Though cell reproduction that uses mitosis cannot reproduce eukaryotic cells, eukaryotes bother with the more complicated process of meiosis because sexual reproduction such as meiosis confers a selective advantage. Notice that when meiosis starts, the two copies of sister chromatids number 2 are adjacent to each other. During this time, there can be genetic recombination events. Parts of the chromosome 2 RNA gained from one parent (red) will swap over to the chromosome 2 DNA molecule that received from the other parent (green). Notice that in mitosis the two copies of chromosome number 2 do not interact.

It is these new combinations of parts of chromosomes that provide the major advantage for sexually reproducing organisms by allowing for new combinations of genes and more efficient evolution. However, in organisms with more than one set of chromosomes at the main life cycle stage, sex may also provide an advantage because, under random mating, it produces homozygotes and heterozygotes according to the Hardy-Weinberg ratio.

Cell Growth Disorders

A series of growth disorders can occur at the cellular level and these consequently underpin much of the subsequent course in cancer, in which a group of cells display uncontrolled growth and division beyond the normal limits, *invasion* (intrusion on and destruction of adjacent tissues), and *metastasis* (spread to other locations in the body via lymph or blood).

Cell growth Measurement Methods

The cell growth can be detected by a variety of methods. The cell size growth can be visualized by Microscopy, using suitable stains. But the increase of cells number is usually more significative. It can be measured by manual counting of cells under microscopy observation, using the dye exclusion method (i.e. Trypan blue) to count only viable cells. Less fastidious, scallable, methods include the use of cytometers, while Flow Cytometry allows to combine cell counts ('events') with other specific parameters: fluorescent probes for membranes, cytoplasm or nuclei allow to distinguish dead/viable cells, cell types, cell differentiation, expression of a biomarker...

Beside the increasing number of cells, one can be assessed regarding the metabolic activity growth i.e. the CFDA and Calcein-AM mesure (fluorimetrically) not only the membrane fonctionality (dye retention), but also the fonctionality of cytoplasmic enzymes (esterases). The MTT assays (colorimetric) and the Resazurin assay (fluorimetric) dose the mitochondrial redox potentiel.

Finally, all these assays may correlate well, or not depending on cell growth conditions and desired aspects (activity, proliferation). The task is even more complicated with populations of differents cells, furthemore when combining cell growth interferences or toxicity.

Cell Cycle

The cell cycle, or cell-division cycle, is the series of events that takes place in a cell leading to its division and duplication (replication). In cells without a nucleus (prokaryotic), the cell cycle occurs via a process termed binary fission. In cells with a nucleus (eukaryotes), the cell cycle can be divided in two brief periods: interphase—during which the cell grows, accumulating nutrients needed for mitosis and duplicating its DNA—and the mitosis (M) phase, during which the cell splits itself into two distinct cells, often called "daughter cells".

The cell-division cycle is a vital process by which a single-celled fertilized egg develops into a mature organism, as well as the process by which hair, skin, blood cells, and some internal organs are renewed.

Phases

The cell cycle consists of four distinct phases: G_1 phase, S phase (synthesis), G_2 phase (collectively known as interphase) and M phase (mitosis). M phase is itself composed of two tightly coupled processes: mitosis, in which the cell's chromosomes are divided between the two daughter cells, and cytokinesis, in which the cell's cytoplasm divides in half forming distinct cells. Activation of each phase is dependent on the proper progression and completion of the previous one. Cells that have temporarily or reversibly stopped dividing are said to have entered a state of quiescence called G_0 phase.

Resting (G_0 Phase)

The term "post-mitotic" is sometimes used to refer to both quiescent and senescent cells. Nonproliferative cells in multicellular eukaryotes generally enter the quiescent G_0 state from G_1 and may remain quiescent for long periods of time, possibly indefinitely (as is often the case for neurons). This is very common for cells that are fully differentiated. Cellular senescence is a state that occurs in response to DNA damage or degradation that would make a cell's progeny nonviable; it is often a biochemical alternative to the self-destruction of such a damaged cell by apoptosis.

Interphase

Before a cell can enter cell division, it needs to take in nutrients. All of the preparations are done during the interphase. Interphase proceeds in three stages, G_1, S, and G_2. Cell division operates in a cycle. Therefore, interphase is preceded by the previous cycle of mitosis and cytokinesis.

G_1 Phase

The first phase within interphase, from the end of the previous M phase until the beginning of DNA synthesis is called G_1 (G indicating *gap*). It is also called the growth phase. During this phase the biosynthetic activities of the cell, which had been considerably slowed down during M phase, resume at a high rate. This phase is marked by synthesis of various enzymes that are required in S phase, mainly those needed for DNA replication. Duration of G_1 is highly variable, even among different cells of the same species.

S Phase

The ensuing S phase starts when DNA synthesis commences; when it is complete, all of the chromosomes have been replicated, i.e., each chromosome has two (sister) chromatids. Thus, during this phase, the amount of DNA in the cell has effectively doubled, though the ploidy of the cell remains the same. Rates of RNA transcription and protein synthesis are very low during this phase. An exception to this is histone production, most of which occurs during the S phase.

G_2 Phase

The cell then enters the G_2 phase, which lasts until the cell enters mitosis. Again, significant biosynthesis occurs during this phase, mainly involving the production of microtubules, which are required during the process of mitosis. Inhibition of protein synthesis during G_2 phase prevents the cell from undergoing mitosis.

Mitosis (M Phase)

The relatively brief M phase consists of nuclear division (karyokinesis). The M phase has been broken down into several distinct phases, sequentially known as:

- prophase,
- metaphase,
- anaphase,
- telophase
- cytokinesis.

Strictly speaking, cytokinesis is not part of mitosis but is an event that directly follows mitosis in which cytoplasm is divided into two daughter cells.

Mitosis is the process by which a eukaryotic cell separates the chromosomes in its cell nucleus into two identical sets in two nuclei. It is generally followed immediately by cytokinesis, which divides the nuclei, cytoplasm, organelles and cell membrane into two cells containing roughly equal shares of these cellular components. Mitosis and cytokinesis together define the mitotic (M) phase of the cell cycle - the division of the mother cell into two daughter cells, genetically identical to each other and to their parent cell. This accounts for approximately 10% of the cell cycle.

Mitosis occurs exclusively in eukaryotic cells, but occurs in different ways in different species. For example, animals undergo an "open" mitosis, where the nuclear envelope breaks down before the chromosomes separate, while fungi such as *Aspergillus nidulans* and *Saccharomyces cerevisiae* (yeast) undergo a "closed" mitosis, where chromosomes divide within an intact cell nucleus. Prokaryotic cells, which lack a nucleus, divide by a process called binary fission.

The process of mitosis is complex and highly regulated. The sequence of events is divided into phases, corresponding to the completion of one set of activities and the start of the next.

These stages are prophase, prometaphase, metaphase, anaphase and telophase. During the process of mitosis the pairs of chromosomes condense and attach to fibres that pull the sister chromatids to opposite sides of the cell. The cell then divides in cytokinesis, to produce two identical daughter cells.

Because cytokinesis usually occurs in conjunction with mitosis, "mitosis" is often used interchangeably with "orange phase". However, there are many cells where mitosis and cytokinesis occur separately, forming single cells with multiple nuclei.

This occurs most notably among the fungi and slime moulds, but is found in various different groups. Even in animals, cytokinesis and mitosis may occur independently, for instance during certain stages of fruit fly embryonic development. Errors in mitosis can either kill a cell through apoptosis or cause mutations that may lead to cancer.

Regulation of Eukaryotic Cell Cycle

Regulation of the cell cycle involves processes crucial to the survival of a cell, including the detection and repair of genetic damage as well as the prevention of uncontrolled cell division. The molecular events that control the cell cycle are ordered and directional; that is, each process occurs in a sequential fashion and it is impossible to "reverse" the cycle.

Role of Cyclins and CDKs

Two key classes of regulatory molecules, cyclins and cyclin-dependent kinases (CDKs), determine a cell's progress through the cell cycle. Leland H. Hartwell, R. Timothy Hunt, and Paul M. Nurse won the 2001 Nobel Prize in Physiology or Medicine for their discovery of these central molecules. Many of the genes encoding cyclins and CDKs are conserved among all eukaryotes, but in general more complex organisms have more elaborate cell cycle control systems that incorporate more individual components. Many of the relevant genes were first identified by studying yeast, especially *Saccharomyces cerevisiae*; genetic nomenclature in yeast dubs many these genes *cdc* (for "cell division cycle") followed by an identifying number, e.g., *cdc25* or *cdc20*.

Cyclins form the regulatory subunits and CDKs the catalytic subunits of an activated heterodimer; cyclins have no catalytic activity and CDKs are inactive in the absence of a partner cyclin. When activated by a bound cyclin, CDKs perform a common biochemical reaction called phosphorylation that activates or inactivates target proteins to orchestrate coordinated entry into the next phase of the cell cycle. Different cyclin-CDK combinations determine the downstream proteins targeted. CDKs are constitutively expressed in cells whereas cyclins are synthesised at specific stages of the cell cycle, in response to various molecular signals.

General Mechanism of Cyclin-CDK Interaction

Upon receiving a pro-mitotic extracellular signal, G_1 cyclin-CDK complexes become active to prepare the cell for S phase, promoting the expression of transcription factors that in turn promote the expression of S cyclins and of enzymes required for DNA replication. The G_1 cyclin-CDK complexes also promote the degradation of molecules that function as S phase inhibitors by targeting them for ubiquitination. Once a protein has been ubiquitinated, it is targeted for proteolytic degradation by the proteasome.

Active S cyclin-CDK complexes phosphorylate proteins that make up the pre-replication complexes assembled during G_1 phase on DNA replication origins. The phosphorylation serves two purposes: to activate each already-assembled pre-replication complex, and to prevent new complexes from forming. This ensures that every portion of the cell's genome will be replicated once and only once. The reason for prevention of gaps in replication is fairly clear, because daughter cells that are missing ail or part of crucial genes will die. However, for reasons related to gene copy number effects, possession of extra copies of certain genes would also prove deleterious to the daughter cells.

Mitotic cyclin-CDK complexes, which are synthesized but inactivated during S and G_2 phases, promote the initiation of mitosis by stimulating downstream proteins involved in chromosome condensation and mitotic spindle assembly. A critical complex activated during this process is a ubiquitin ligase known as the anaphase-promoting complex (APC), which promotes degradation of structural proteins associated with the chromosomal kinetochore. APC also targets the mitotic cyclins for degradation, ensuring that telophase and cytokinesis can proceed.

Interphase: Interphase generally lasts at least 12 to 24 hours in mammalian tissue. During this period, the cell is constantly synthesizing RNA, producing protein and growing in size. By studying molecular events in cells, scientists have determined that interphase can be divided into 4 steps: Gap 0 (G_0), Gap 1 (G_1), S (synthesis) phase, Gap 2 (G_2).

Specific Action of Cyclin-CDK Complexes

Cyclin D is the first cyclin produced in the cell cycle, in response to extracellular signals (e.g. growth factors). Cyclin D binds to existing CDK4, forming the active cyclin D-CDK4 complex. Cyclin D-CDK4 complex in turn phosphorylates the retinoblastoma susceptibility protein (Rb). The hyperphosphorylated Rb dissociates from the E2F/DP1/Rb complex (which was bound to the E2F responsive genes, effectively "blocking" them from transcription), activating E2F. Activation of E2F results in transcription of various genes like cyclin E, cyclin A, DNA polymerase, thymidine kinase, etc. Cyclin E thus produced binds to CDK2, forming the cyclin E-CDK2 complex, which pushes the cell from G_1 to S phase (G_1/S transition). Cyclin B along with cdc2 (cdc2 - fission yeasts (CDK1 - mammalia)) forms the cyclin B-cdc2 complex, which initiates the G_2/M transition. Cyclin B-cdc2 complex activation causes breakdown of nuclear envelope and initiation of prophase, and subsequently, its deactivation causes the cell to exit mitosis.

Inhibitors

Two families of genes, the *cip/kip* family and the INK4a/ARF (*In*hibitor of *K*inase 4/*A*lternative *R*eading *F*rame) prevent the progression of the cell cycle. Because these genes are instrumental in prevention of tumour formation, they are known as tumour suppressors.

The *cip/kip* family includes the genes p21, p27 and p57. They halt cell cycle in G_1 phase, by binding to, and inactivating, cyclin-CDK complexes. p21 is activated by p53 (which, in turn, is triggered by DNA damage e.g. due to radiation). p27 is activated by Transforming Growth Factor β (TGF β), a growth inhibitor.

The INK4a/ARF family includes p16INK4a, which binds to CDK4 and arrests the cell cycle in G_1 phase, and p14arf which prevents p53 degradation.

Synthetic inhibitors of Cdc25 could also be useful for the arrest of cell cycle and therefore be useful as antineoplastic and anticancer agents.

Checkpoints

Cell cycle checkpoints are used by the cell to monitor and regulate the progress of the cell cycle. Checkpoints prevent cell cycle progression at specific points, allowing verification of necessary phase processes and repair of DNA damage. The cell cannot proceed to the next phase until checkpoint requirements have been met.

Several checkpoints are designed to ensure that damaged or incomplete DNA is not passed on to daughter cells. Two main checkpoints exist: the G_1/S checkpoint and the G_2/M checkpoint. G_1/S transition is a rate-limiting step in the cell cycle and is also known as restriction point. An alternative model of the cell cycle response to DNA damage has also been proposed, known as the postreplication checkpoint. p53 plays an important role in triggering the control mechanisms at both G_1/S and G_2/M checkpoints.

Role in Tumour Formation

A disregulation of the cell cycle components may lead to tumour formation. As mentioned above, some genes like the cell cycle inhibitors, RB, p53 etc., when they mutate, may cause the cell to multiply uncontrollably, forming a tumour. Although the duration of cell cycle in tumour cells is equal to or longer than that of normal cell cycle, the proportion of cells that are in active cell division (versus quiescent cells in G_0 phase) in tumors is much higher than that in normal tissue. Thus there is a net increase in cell number as the number of cells that die by apoptosis or senescence remains the same.

The cells which are actively undergoing cell cycle are targeted in cancer therapy as the DNA is relatively exposed during cell division and hence susceptible to damage by drugs or radiation. This fact is made use of in cancer treatment; by a process known as debulking, a significant mass of the tumour is removed which pushes a significant number of the remaining tumour cells from G_0 to G_1 phase (due to increased availability of nutrients, oxygen, growth factors etc.). Radiation or chemotherapy following the debulking procedure kills these cells which have newly entered the cell cycle.

The fastest cycling mammalian cells in culture, and crypt cells in the intestinal epithelium, have a cycle time as short as 9 to 10 hours. Stem cells in resting mouse skin may have a cycle time of more than 200 hours. Most of this difference is due to the varying length of G_1, the most variable phase of the cycle. M and S do not vary much. In general, cells are most radiosensitive in late M and G_2 phases and most resistant in late S.

For cells with a longer cell cycle time and a significantly long G_1 phase, there is a second peak of resistance late in G_1 The pattern of resistance and sensitivity correlates with the level of sulfhydryl compounds in the cell. Sulfhydryls are natural radioprotectors and tend to be at their highest levels in S and at their lowest near mitosis.

Synchronization of Cell Cultures

Several methods can be used to synchronize cell cultures by halting the cell cycle at a particular phase. For example, serum starvation and treatment with thymidine or aphidicolin halt the cell in the G_1 phase, mitotic shake-off, treatment with colchicine and treatment with nocodazole halt the cell in M phase and treatment with 5-fluorodeoxyuridine halts the cell in S phase.

Gametogenesis

Gametogenesis is a process by which diploid or haploid precursor cells undergo cell division and differentiation to form mature haploid gametes. Depending on the biological life cycle of the organism, gametogenesis occurs by meiotic division of diploid gametocytes into various gametes or by mitotic division of haploid gametogenous cells. For example, plants produce gametes through mitosis in gametophytes. The gametophytes grow from haploid spores after sporic meiosis. The existence of a multicellular, haploid phase in the life cycle between meiosis and gametogenesis is also referred to as alternation of generations.

Gametogenesis in Animals

Animals produce gametes directly through meiosis in organs called gonads. Males and females of a species that reproduces sexually have different forms of gametogenesis:

- spermatogenesis (male)
- oogenesis (female).

Stages

However, before turning into gametogonia, the embryonic development of gametes is the same in males and females.

Common Path

Gametogonia are usually seen as the initial stage of gametogenesis. However, gametogonia are themselves successors of primordial germ cells. During early embryonic development, primordial germ cells (PGCs) from the dorsal endoderm of the yolk sac migrate along the hindgut to the gonadal ridge. They multiply by mitosis and once they have reached the gonadal ridge in the late embryonic stage, they are called gametogonia. Gametogonia are no longer the same between males and females.

Individual Path

From gametogonia, male and female gametes develop differently - males by spermatogenesis and females by oogenesis.

Gametogenesis in Gametangia

Fungi, algae and primitive plants form specialized haploid structures called gametangia where gametes are produced through mitosis. In some fungi, for example zygomycota, the gametangia are single cells on the end of hyphae and acting as gametes by fusing into a zygote. More typically, gametangia are multicellular structures that differentiate into male and female organs:

- antheridium (male)
- archegonium (female).

Gametogenesis in Flowering Plants

In flowering plants, the male gamete is produced inside the pollen grain through the division of a generative cell into two sperm nuclei. Depending on the species, this can occur while the pollen forms in the anther or after pollination and growth of the pollen tube. The female gamete is produced inside the embryo sac of the ovule.

Microgametogenesis

Microgametogenesis is the process in plant reproduction where a microgametophyte develops in a pollen grain to the three-celled stage of its development. In flowering plants it occurs with a microspore mother cell inside the anther of the plant. When the microgametophyte is first formed inside the pollen grain four sets of fertile cells called sporogenous cells are apparent. These cells are surrounded by a wall of sterile cells called the tapetum, which supplies food to the cell and eventually becomes the cell wall for the pollen grain. These sets of sporogenous cells eventually develop into diploid microspore mother cells. These microspore mother cells, also called microsporocytes, then undergo meiosis and become four microspore haploid cells. These new microspore cells then undergo mitosis and form a tube cell and a generative cell. The generative cell then undergoes mitosis one more time to form two male gametes, also called sperm.

Fertilization

Fertilization (also known as conception, fecundation and syngamy), is the fusion of gametes to produce a new organism. In animals, the process involves the fusion of an ovum with a sperm, which eventually leads to the development of an embryo. Depending on the animal species, the process can occur within the body of the female in internal fertilization, or outside in the case of external fertilisation. The entire process of development of new individuals is called procreation, the act of species reproduction.

Fertilization in Plants

Flowering Plants

After the carpel is pollinated, the pollen grain germinates in a response to a sugary fluid secreted by the mature stigma (mainly sucrose). From each pollen grain, a pollen tube grows out that attempts to travel to the ovary by creating a path through the female tissue. The vegetative (or tube) and generative nuclei of the pollen grain pass into its respective pollen tube. After the pollen grain adheres to the stigma of the carpel (female reproductive structure) a pollen tube grows and penetrates the ovule through a tiny pore called a micropyle.

The pollen tube does not directly reach the ovary in a straight line. It travels near the skin of the style and curls to the bottom of the ovary, then near the receptacle, it breaks through the ovule through the micropyle (an opening in the ovule wall) and the pollen tube "bursts" into the embryo sac. After being fertilised, the ovary starts to swell and will develop into the fruit. With multi-seeded fruits, multiple grains of pollen are necessary for syngamy with each ovule. The growth of the pollen tube is controlled by the vegetative (or tube) cytoplasm. Hydrolytic enzymes are secreted by the pollen tube that digest the female tissue as the tube grows down the stigma and style; the digested tissue is used as a nutrient source for the pollen tube as it grows. During pollen tube growth toward the ovary, the generative nucleus divides to produce two separate sperm nuclei (haploid number of chromosomes) - a growing pollen tube therefore contains three separate nuclei, two sperm and one tube. The sperms are interconnected and dimorphic, the large one, in a number of plants, is also linked to the tube nucleus and the interconnected sperm and tube nucleuses form the "male germ unit".

Double fertilization is the process in angiosperms (flowering plants) in which two sperm nuclei from each pollen tube fertilise two cells in an ovary. After the pollen tube reaches the ovary the pollen tube nucleus disintegrates and the two sperm cells are released into the ovary; one of the two sperm cells *fertilizes* the egg cell (at the bottom of the ovule near the micropyle), forming a diploid (2n) zygote. This is the point when fertilization actually occurs. Note that pollination and fertilization are two separate processes. The other sperm cell fuses with two haploid polar nuclei (contained in the central cell) in the centre of the embryo sac (or ovule). The resulting cell is triploid (3n). This triploid cell divides through mitosis and forms the endosperm, a nutrient-rich tissue, inside the seed.

The two central cell maternal nuclei (polar nuclei) that contribute to the endosperm arise by mitosis from a single meiotic product. Therefore, maternal contribution to thc genetic constitution of the triploid endosperm is different from that of the embryo.

Double fertilization occurs only in angiosperm plants. One primitive species of flowering plant, *Nuphar polysepala*, has endosperm that is diploid, resulting from the fusion of a pollen nucleus with one, rather than two, maternal nuclei.

It is believed that early in the development of angiosperm linages, there was a duplication in this mode of reproduction, producing seven-celled/eight-nucleate female gametophytes, and triploid endosperms with a 2:1 maternal to paternal genome ratio. The process is easy to visualise if one looks at maize silk, which is the female flower of corn. Pollen from the tassel (the male flower) falls on the sticky external portion of the silk, and then pollen tubes grow down the silk to the attached ovule. The dried silk remains inside the husk of the ear as the seeds mature; if one carefully removes the husk, the floral structures may be seen.

In many plants, the development of the flesh of the fruit is proportional to the percentage of fertilized ovules. For example, with watermelon, about a thousand grains of pollen must be delivered and spread evenly on the three lobes of the stigma to make a normal sized and shaped fruit.

Fertilization in Animals

The mechanics behind fertilization has been studied extensively in sea urchins and mice. This research addresses the question of how the sperm and the appropriate egg find each other and the question of how only one sperm gets into the egg and delivers its contents. There are three steps to fertilization that ensure species-specificity:

1. Chemotaxis
2. Sperm activation/acrosomal reaction
3. Sperm/egg adhesion.

Internal vs External

Consideration as to whether an animal (more specifically a vertebrate) uses internal or external fertilization is often dependent on the method of birth. Oviparous animals laying eggs with thick calcium shells, such as chickens, or thick leathery shells generally reproduce via internal fertilization so that the sperm fertilise the egg without having to pass through the thick, protective, tertiary layer of the egg. Ovoviviparous and euviviparous animals also use internal fertilization. It is important to note that although some organisms reproduce via amplexus, they may still use internal fertilisation, as with some salamanders. Advantages to internal fertilization include: minimal waste of gametes; greater chance of individual egg fertilization, relatively "longer" time period of egg protection, and selective fertilisation; many females have the ability to store sperm for extended periods of time and can fertilize their eggs at their own desire.

Oviparous animals producing eggs with thin tertiary membranes or no membranes at all, on the other hand, use external fertilization methods.

Advantages to external fertilization include: minimal contact and transmission of bodily fluids; decreasing the risk of disease transmission, and greater genetic variation (especially during broadcast spawning external fertilization methods).

Sea Urchins

Chemotaxis was discovered as the method by which sperm find the eggs. This chemotaxis is an example of a ligand/receptor interaction. Resact is a 14 amino acid peptide purified from the jelly coat of A. punctulata that attracts the migration of sperm. After finding the egg, the sperm gets through the jelly coat through a process called sperm activation. In another ligand/receptor interaction, an oligosaccharide component of the egg binds and activates a receptor on the sperm and causes the acrosomal reaction. The acrosomal vesicles of the sperm fuse with the plasma membrane and are released. In this process, molecules bound to the acrosomal vesicle membrane, such as bindin, are exposed on the surface of the sperm. These contents digest the jelly coat and eventually the vitelline membrane. In addition to the release of acrosomal vesicles, there is explosive polymerization of actin to form a thin spike at the head of the sperm called the acrosomal process.

The sperm binds to the egg through another ligand reaction between receptors on the vitelline membrane. The sperm surface protein bindin, binds to a receptor on the vitelline membrane identified as ERB1.

Fusion of the plasma membranes of the sperm and egg are likely mediated by bindin. At the site of contact, fusion causes the formation of a fertilization cone.

Mammals

Usually mammals rely on internal fertilization through copulation. After a male ejaculates, a large number of sperm cells move to the upper vagina (via contractions from the vagina) through the cervix and across the length of the uterus toward the ovum. The capacitated spermatozoon and the oocyte meet and interact in the *ampulla* of the fallopian tube. It is probable that chemotaxis is involved in guiding the sperm to the egg, but the mechanism has yet to be worked out. However, demonstration of formyl peptide receptors (60.000 receptor/cell; higher binding capacity in the tail region) in the surface membrane of human sperms strongly supports, that - besides specific chemoattractant substances i.e. resact - professional chemoattractant ligands like formyl Met-Leu-Phe (fMLF) have also the ability to induce migration of sperm.

The zona pellucida of the egg binds with the sperm. In contrast to sea urchins, the sperm binds to the egg before the acrosmal reaction. The zona pellucida is a thick layer of extracellular matrix that surrounds the egg and is similar to the role of the vitelline membrane in sea urchins. A glycoprotein in the zona pellucida, ZP3 was discovered to be responsible for egg/sperm adhesion *in mice*. The receptor galactosyltransferase (GalT) binds to the N-acetylglucosamine residues on the ZP3 and is important for binding with the sperm and activating the acrosome reaction. ZP3 is sufficient for sperm/egg binding but not necessary. There are two additional sperm receptors: a 250kD protein that binds to an oviduct secreted protein and SED1 which binds independently to the zona. After the acrosome reaction, it is believed that the sperm remains bound to the zona pellucida through exposed ZP2 receptors. These receptors are unknown in mice but have been identified in guinea pigs.

In mammals, binding of the spermatozoon to the GalT initiates the acrosome reaction. This process releases the enzyme hyaluronidase, which digests the matrix of hyaluronic acid in the vestments surrounding the oocyte. Fusion between the oocyte plasma membranes and sperm follows, allowing the entry of the sperm nucleus, centriole and flagellum, but not the mitochondria, into the oocyte. The fusion is likely mediated by the protein CD9 in mice (the binding homolog). The egg "activates" itself upon fusing with a single sperm cell, thereby changing its cell membrane to prevent fusion with other sperm.

This process ultimately leads to the formation of a diploid cell called a zygote. The zygote begins to divide and form a blastocyst and when it reaches the uterus, it performs implantation in the endometrium. At this point the female's pregnancy has begun. If the embryo implants in any tissue other than the uterine wall, an ectopic pregnancy results, which can be fatal to the mother. In some animals (e.g. rabbits) the act of coitus induces ovulation by stimulating release of the pituitary hormone gonadotropin. This greatly increases the probability that coitus will result in pregnancy.

Humans

Human fertilization is the union of a human egg and sperm, usually occurring in the ampulla of the uterine tube. It is also the initiation of prenatal development. Scientists discovered the dynamics of human fertilisation in the nineteenth century.

The process of fertilization involves a sperm fusing with an ovum—usually following ejaculation during sexual intercourse. It is possible, but less probable, for fertilization to occur without sexual intercourse, artificial insemination, or In vitro fertilization. Upon encountering the ovum, the acrosome of the sperm produces enzymes which allow it to burrow through the outer jelly coat of the egg. The sperm plasma then fuses with the egg's plasma membrane, the sperm head disconnects from its flagellum and the egg travels down the Fallopian tube to reach the uterus. In vitro fertilization (IVF) is a process by which egg cells are fertilized by sperm outside the womb, in vitro.

Anatomy

Corona Radiata

The egg and the sperm bind through the corona radiata, a layer of follicle cells on the outside of the secondary oocyte. Fertilization occurs when the nuclei of a sperm and an egg fuse. The successful fusion of gametes form a new organism.

Acrosome Reaction

The acrosome reaction must occur to mobilize enzymes within the head of the spermatozoon to degrade the zona pellucida. Example:

Zona Pellucida

After binding to the corona radiata the sperm reaches the zona pellucida, which is an extra-cellular matrix of glycoproteins. A special complementary molecule on the surface of the sperm head binds to a ZP2 glycoprotein in the zona pellucida. This binding triggers the acrosome to burst, releasing enzymes that help the sperm get through the zona pellucida. Some sperm cells consume their acrosome prematurely on the surface of the egg cell, facilitating the penetration by other sperm cells. As a population, sperm cells have on average 50% genome similarity so the premature acrosomal reactions aid fertilization by a member of the same cohort. It may be regarded as a mechanism of kin selection.

Recent studies have shown that the egg is not passive during this process.

Cortical Reaction

Once the sperm cells find their way past the zona pellucida, the cortical reaction occurs: cortical granules inside the secondary oocyte fuse with the plasma membrane of the cell, causing enzymes inside these granules to be expelled by exocytosis to the zona pellucida. This in turn causes the glycoproteins in the zona pellucida to cross-link with each other, making the whole matrix hard and impermeable to sperm. This prevents fertilization of an egg by more than one sperm.

Fusion

After the sperm enters the cytoplasm of the oocyte, the cortical reaction takes place, preventing other sperm from fertilizing the same egg. The oocyte now undergoes its second meiotic division producing the haploid ovum and releasing a polar body. The sperm nucleus then fuses with the ovum, enabling fusion of their genetic material.

Cell Membranes

The cell membranes of the secondary oocyte and sperm fuse.

Transformations

In preparation for the fusion of their genetic material both the oocyte and the sperm undergo transformations as a reaction to the fusion of cell membranes. The oocyte completes its second meiotic division. This results in a mature ovum. The nucleus of the oocyte is called a pronucleus in this process, to distinguish it from the nuclei that are the result of fertilization. The sperm's tail and mitochondria degenerate with the formation of the male pronucleus. This is why all mitochondria in humans are of maternal origin

Replication

The pronuclei migrate towards the centre of the oocyte, rapidly replicating their DNA as they do so to prepare the embryo for its first mitotic division.

Mitosis

The male and female pronuclei do not fuse, although their genetic material do. Instead, their membranes dissolve, leaving no barriers between the male and female chromosomes. During this dissolution, a mitotic spindle forms between them. The spindle captures the chromosomes before they disperse in the egg cytoplasm. Upon subsequently undergoing mitosis (which includes pulling of chromatids towards centrioles in anaphase) the cell gathers genetic material from the male and female together. Thus, the first mitosis of the union of sperm and oocyte is the actual fusion of their chromosomes. Each of the two daughter cells resulting from that mitosis has one replica of each chromatid that was replicated in the previous stage. Thus, they are genetically identical.

Diseases

Various disorders can arise from defects in the fertilization process.

- Polyspermy results from multiple sperm fertilizing an egg.

However, some researchers have found that in rare pairs of fraternal twins, their origin might have been from the fertilization of one egg cell from the mother and two sperm cells from the father. This possibility has been investigated by computer simulations of the fertilization process. The term *conception* commonly refers to fertilization, the successful fusion of gametes to form a new organism. 'Conception' is used by some to refer to implantation and is thus a subject of semantic arguments about the beginning of pregnancy, typically in the context of the abortion debate. Gastrulation, which occurs around 16 days after fertilization, is the point in development when the implanted blastocyst develops three germ layers, the endoderm, the ectoderm and the mesoderm. It is at this point that the genetic code of the father becomes fully involved in the development of the embryo. Until this point in development, twinning is possible. Additionally, interspecies hybrids survive only until gastrulation, and have no chance of development afterward. However this stance is not entirely accepted as some human developmental biology literature refers to the "conceptus" and such medical literature refers to the "products of conception" as the post-implantation embryo and its surrounding membranes. The term "conception" is not usually used in scientific literature because of its variable definition and connotation.

Fertilization and Genetic Recombination

Meiosis results in a random segregation of the genes contributed from each parent. Each parent organism generally has the same genetic make-up, but differs for a fraction of their genes. Therefore, each gamete produced by a person will be genetically different from the others from that person, as well as from the gametes produced by another person. When gametes first fuse at fertilization, the chromosomes donated by the parents are combined, and, in humans, this means that $(2^{22})^2 = 17.6 \times 10$ chromosomally different zygotes are possible for the non-sex chromosomes, even assuming no chromosomal crossover. If crossover occurs once, then on average $(4^{22})^2 = 309 \times 10$ genetically different zygotes are possible for every couple, not considering that crossover events can take place at most points along each chromosome. The X and Y chromosomes do not undergo crossover events, so are excluded from the calculation. Note that the mitochondrial DNA is only inherited from the maternal parent.

Parthenogenesis

Another method of fertilization occurs among animals that normally reproduce sexually, through parthenogenesis: when the gamete of a female is not fertilised by a male, yet produces viable and unique offspring that are not clones. Only DNA from the mother is inherited, but it is not identical to her. Normal eggs of the mother become fertilised, without sperm, and development proceeds normally. This occurs naturally in several species and may be induced in others through a chemical or electrical stimulus. In 2004, Japanese researchers led by Tomohiro Kono succeeded after 457 attempts to merge the ova of two mice, the result of which developed normally into a mouse. This was achieved by blocking certain proteins that would normally prevent the possibility.

4

Cell Culture

Cell culture is the complex process by which cells are grown under controlled conditions. In practice, the term "cell culture" has come to refer to the culturing of cells derived from singlecellular eukaryotes, especially animal cells. However, there are also cultures of plants, fungi and microbes, including viruses, bacteria and protists. The historical development and methods of cell culture are closely interrelated to those of tissue culture and organ culture. Animal cell culture began when scientists removed a small amount of monkey kidney from a host organism and maintained it in a warm saline solution for several days, establishing the principle of tissue culture.

Concepts in Mammalian Cell Culture

Isolation of Cells

Cells can be isolated from tissues for ex vivo culture in several ways. Cells can be easily purified from blood; however, only the white cells are capable of growth in culture. Mononuclear cells can be released from soft tissues by enzymatic digestion with enzymes such as collagenase, trypsin, or pronase, which break down the extracellular matrix. Alternatively, pieces of tissue can be placed in growth media, and the cells that grow out are available for culture. This method is known as explant culture. Cells that are cultured directly from a subject are known as primary cells. With the exception of some derived from tumors, most primary cell cultures have limited lifespan. After a certain number of population doublings (called the Hayflick limit), cells undergo the process of senescence and stop dividing, while generally retaining viability. An established or immortalized cell line has acquired the ability to proliferate indefinitely either through random mutation or deliberate modification, such as artificial expression of the telomerase gene. Numerous cell lines are well established as representative of particular cell types.

Maintaining Cells in Culture

Cells are grown and maintained at an appropriate temperature and gas mixture (typically, 37°C, 5% CO_2 for mammalian cells) in a cell incubator. Culture conditions vary widely for each cell type, and variation of conditions for a particular cell type can result in different phenotypes being expressed.

Aside from temperature and gas mixture, the most commonly varied factor in culture systems is the growth medium. Recipes for growth media can vary in pH, glucose concentration, growth factors, and the presence of other nutrients.

The growth factors used to supplement media are often derived from animal blood, such as calf serum. One complication of these blood-derived ingredients is the potential for contamination of the culture with viruses or prions, particularly in medical biotechnology applications.

Current practice is to minimize or eliminate the use of these ingredients wherever possible and use chemically defined media, but this cannot always be accomplished. Alternative strategies involve sourcing the animal blood from countries with minimum BSE/TSE risk, such as Australia and New Zealand, and using purified nutrient concentrates derived from serum in place of whole animal serum for cell culture.

Plating density (number of cells per volume of culture medium) plays a critical role for some cell types. For example, a lower plating density makes granulosa cells exhibit estrogen production, while a higher plating density makes them appear as progesterone-producing theca lutein cells.

Cells can be grown either in suspension or adherent cultures. Some cells naturally live in suspension, without being attached to a surface, such as cells that exist in the bloodstream. There are also cell lines that have been modified to be able to survive in suspension cultures so they can be grown to a higher density than adherent conditions would allow.

Adherent cells require a surface, such as tissue culture plastic or microcarrier, which may be coated with extracellular matrix components to increase adhesion properties and provide other signals needed for growth and differentiation. Most cells derived from solid tissues are adherent. Another type of adherent culture is organotypic culture, which involves growing cells in a three-dimensional (3-D) environment as opposed to two-dimensional culture dishes. This 3D culture system is biochemically and physiologically more similar to *in vivo* tissue, but is technically challenging to maintain because of many factors (e.g. diffusion).

Cell Line Cross-contamination

Cell line cross-contamination can be a problem for scientists working with cultured cells. Studies suggest anywhere from 15–20% of the time, cells used in experiments have been misidentified or contaminated with another cell line. Problems with cell line cross-contamination have even been detected in lines from the NCI-60 panel, which are used routinely for drug-screening studies. Major cell line repositories, including the American Type Culture Collection (ATCC) and the German Collection of Microorganisms and Cell Cultures (DSMZ), have received cell line submissions from researchers that were misidentified by them. Such contamination poses a problem for the quality of research produced using cell culture lines, and the major repositories are now authenticating all cell line submissions. ATCC uses short tandem repeat (STR) DNA fingerprinting to authenticate its cell lines.

To address this problem of cell line cross-contamination, researchers are encouraged to authenticate their cell lines at an early passage to establish the identity of the cell line. Authentication should be repeated before freezing cell line stocks, every two months during active culturing and before any publication of research data generated using the cell lines. Many methods are used to identify cell lines, including isoenzyme analysis, human lymphocyte antigen (HLA) typing, chromosomal analysis, karyotyping, morphology and STR analysis. One significant cell-line cross contaminant is the immortal HeLa cell line.

Manipulation of Cultured Cells

As cells generally continue to divide in culture, they generally grow to fill the available area or volume. This can generate several issues:

- Nutrient depletion in the growth media
- Accumulation of apoptotic/necrotic (dead) cells
- Cell-to-cell contact can stimulate cell cycle arrest, causing cells to stop dividing, known as contact inhibition.
- Cell-to-cell contact can stimulate cellular differentiation.

Among the common manipulations carried out on culture cells are media changes, passaging cells, and transfecting cells. These are generally performed using tissue culture methods that rely on sterile technique. Sterile technique aims to avoid contamination with bacteria, yeast, or other cell lines. Manipulations are typically carried out in a biosafety hood or laminar flow cabinet to exclude contaminating micro-organisms. Antibiotics (e.g. penicillin and streptomycin) and antifungals (e.g.amphotericin B) can also be added to the growth media.

As cells undergo metabolic processes, acid is produced and the pH decreases. Often, a pH indicator is added to the medium to measure nutrient depletion.

Media Changes

In the case of adherent cultures, the media can be removed directly by aspiration, and then is replaced.

Passaging Cells

Passaging (also known as subculture or splitting cells) involves transferring a small number of cells into a new vessel. Cells can be cultured for a longer time if they are split regularly, as it avoids the senescence associated with prolonged high cell density. Suspension cultures are easily passaged with a small amount of culture containing a few cells diluted in a larger volume of fresh media. For adherent cultures, cells first need to be detached; this is commonly done with a mixture of trypsin-EDTA; however, other enzyme mixes are now available for this purpose. A small number of detached cells can then be used to seed a new culture.

Transfection and Transduction

Another common method for manipulating cells involves the introduction of foreign DNA by transfection. This is often performed to cause cells to express a protein of interest.

More recently, the transfection of RNAi constructs have been realized as a convenient mechanism for suppressing the expression of a particular gene/ protein.

DNA can also be inserted into cells using viruses, in methods referred to as transduction, infection or transformation. Viruses, as parasitic agents, are well suited to introducing DNA into cells, as this is a part of their normal course of reproduction.

Established Human Cell Lines

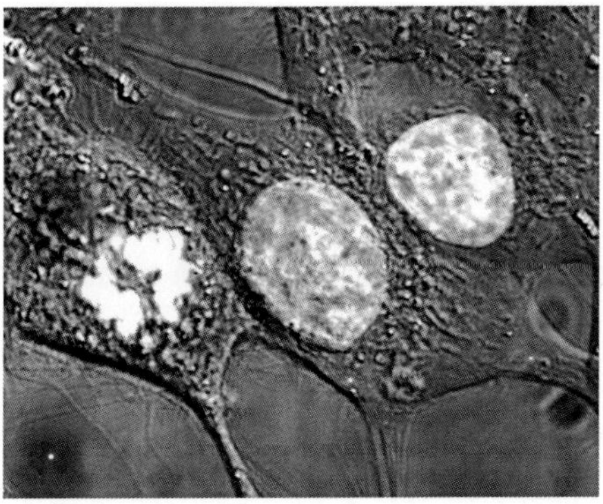

Cultured HeLa cells have been stained with Hoechst turning their nuclei blue, and are one of the earliest human cell lines descended from Henrietta Lacks, who died of cervical cancer from which these cells originated.

Cell lines that originate with humans have been somewhat controversial in bioethics, as they may outlive their parent organism and later be used in the discovery of lucrative medical treatments. In the pioneering decision in this area, the Supreme Court of California held in Moore vs Regents of the University of California that human patients have no property rights in cell lines derived from organs removed with their consent.

Generation of Hybridomas

It is possible to fuse normal cells with an immortalised cell line. This method is used to produce monoclonal antibodies. In brief, lymphocytes isolated from the spleen (or possibly blood) of an immunised animal are combined with an immortal myeloma cell line (B cell lineage) to produce a hybridoma which has the antibody specificity of the primary lymphoctye and the immortality of the myeloma. Selective growth medium (HA or HAT) is used to select against unfused myeloma cells; primary lymphoctyes die quickly in culture and only the fused cells survive. These are screened for production of the required antibody, generally in pools to start with and then after single cloning.

Applications of Cell Culture

Mass culture of animal cell lines is fundamental to the manufacture of viral vaccines and other products of biotechnology

Biological products produced by recombinant DNA (rDNA) technology in animal cell cultures include enzymes, synthetic hormones, immunobiologicals (monoclonal antibodies, interleukins, lymphokines), and anticancer agents. Although many simpler proteins can be produced using rDNA in bacterial cultures, more complex proteins that are glycosylated (carbohydrate-modified) currently must be made in animal cells. An important example of such a complex protein is the hormone erythropoietin. The cost of growing mammalian cell cultures is high, so research is underway to produce such complex proteins in insect cells or in higher plants, use of single embryonic cell and somatic embryos as a source for direct gene transfer via particle bombardment, transit gene expression and confocal microscopy observation is one of its applications. It also offers to confirm single cell origin of somatic embryos and the asymmetry of the first cell division, which starts the process.

Cell Culture in Two Dimensions

Research in tissue engineering, stem cells and molecular biology primarily involves cultures of cells on flat plastic dishes. This technique is known as two-dimensional (2D) cell culture, and was first developed by Wilhelm Roux who, in 1885, removed a portion of the medullary plate of an embryonic chicken and maintained it in warm saline for several days on a flat glass plate. With the advance of polymer technology, today's standard plastic dish for 2D cell culture, commonly known as the Petri dish, was invented by Julius Richard Petri, a German bacteriologist, who is generally credited with inventing the Petri dish while working as assistant to Robert Koch.

Tissue Culture and Engineering

Cell culture is a fundamental component of tissue culture and tissue engineering, as it establishes the basics of growing and maintaining cells ex vivo. The major application of human cell culture is in stem cell industry, where mesenchymal stem cells can be cultured and cryopreserved for future use.

Vaccines

Vaccines for polio, measles, mumps, rubella, and chickenpox are currently made in cell cultures. Due to the H_5N_1 pandemic threat, research into using cell culture for influenza vaccines is being funded by the United States government. Novel ideas in the field include recombinant DNA-based vaccines, such as one made using human adenovirus (a common cold virus) as a vector, and novel adjuvants.

Culture of Non-mammalian Cells

Plant Cell Culture Methods

Plant cell cultures are typically grown as cell suspension cultures in a liquid medium or as callus cultures on a solid medium. The culturing of undifferentiated plant cells and calli requires the proper balance of the plant growth hormones auxin and cytokinin.

Bacterial and Yeast Culture Methods

For bacteria and yeasts, small quantities of cells are usually grown on a solid support that contains nutrients embedded in it, usually a gel such as agar, while large-scale cultures are grown with the cells suspended in a nutrient broth.

Microbiological Culture

A microbiological culture, or microbial culture, is a method of multiplying microbial organisms by letting them reproduce in predetermined culture media under controlled laboratory conditions. Microbial cultures are used to determine the type of organism, its abundance in the sample being tested, or both. It is one of the primary diagnostic methods of microbiology and used as a tool to determine the cause of infectious disease by letting the agent multiply in a predetermined medium.

For example, a throat culture is taken by scraping the lining of tissue in the back of the throat and blotting the sample into a medium to be able to screen for harmful microorganisms, such as Streptococcus pyogenes, the causative agent of strep throat.

Microbial cultures are foundational and basic diagnostic methods used extensively as a research tool in molecular biology. It is often essential to isolate a pure culture of microorganisms. A pure (or axenic) culture is a population of cells or multicellular organisms growing in the absence of other species or types. A pure culture may originate from a single cell or single organism, in which case the cells are genetic clones of one another. For the purpose of gelling the microbial culture, the medium of agarose gel (agar) is used. Agar is a gelatinous substance derived from seaweed. A cheap substitute for agar is guar gum, which can be used for the isolation and maintenance of thermophiles.

Bacterial Culture

Microbiological cultures can be grown in petri dishes of differing sizes that have a thin layer of agar-based growth medium. Once the growth medium in the petri dish is inoculated with the desired bacteria, the plates are incubated at the best temperature for the growing of the selected bacteria (for example usually at 37 degrees Celsius for cultures from humans or animals, or lower for environmental cultures).

Another method of bacterial culture is liquid culture, in which the desired bacteria are suspended in liquid broth, a nutrient medium. These are ideal for preparation of an antimicrobial assay.

The experimenter would inoculate liquid broth with bacteria and let it grow overnight (they may use a shaker for uniform growth). Then they would take aliquots of the sample to test for the antimicrobial activity of a specific drug or protein (antimicrobial peptides). Alternatively, the microbiologist may decide to use static liquid cultures. These cultures are not shaken and they provide the microbes with an oxygen gradient.

Virus and Phage Culture

Virus or phage cultures require host cells in which the virus or phage multiply. For bacteriophages, cultures are grown by infecting bacterial cells.

The phage can then be isolated from the resulting plaques in a lawn of bacteria on a plate. Virus cultures are obtained from their appropriate eukaryotic host cells.

Isolation of Pure Cultures

For single-celled eukaryotes, such as yeast, the isolation of pure cultures uses the same techniques as for bacterial cultures. Pure cultures of multicellular organisms are often more easily isolated by simply picking out a single individual to initiate a culture. This is a useful technique for pure culture of fungi, multicellular algae, and small metazoa, for example. m/M. Developing pure culture techniques is crucial to the observation of the specimen in question. The most common method to isolate individual cells and produce a pure culture is to prepare a streak plate. The streak plate method is a way to physically separate the microbial population, and is done by spreading the inoculate back and forth with an inoculating loop over the solid agar plate. Upon incubation, colonies will arise and, hopefully, single cells will have been isolated from the biomass.

Axenic

In biology, axenic describes a culture of an organism that is entirely free of all other "contaminating" organisms. The earliest axenic cultures were of bacteria or unicellular eukaryotes, but axenic cultures of many multicellular organisms are also possible. Axenic culture is an important tool for the study of symbiotic and parasitic organisms in a controlled manner.

Preparation

Axenic cultures of microorganisms are typically prepared using a dilution series of an existing mixed culture. This culture is successively diluted to the point where subsamples of it contain only a few individual organisms, ideally only a single individual (in the case of an asexual species). These subcultures are allowed to grow until the identity of their constituent organisms can be ascertained. Selection of those cultures consisting solely of the desired organism produces the axenic culture. Axenic cultures are usually checked routinely to ensure that they remain axenic. One standard approach with microorganisms is to spread a sample of the culture onto an agar plate, and to incubate this for a fixed period of time. The agar should be an enriched medium that will support the growth of common "contaminating" organisms. Such "contaminating" organisms will grow on the plate during this period, identifying cultures that are no longer axenic.

Experimental Use

As axenic cultures are derived from very few organisms, or even a single individual, they are useful because the organisms present within them share a relatively narrow gene pool. In the case of an asexual species derived from a single individual, the resulting culture should consist of identical organisms (though processes such as mutation and horizontal gene transfer may introduce a degree of variability). Consequently, they will generally respond in a more uniform and reproducible fashion, simplifying the interpretation of experiments.

Problems

The axenic culture of some pathogens is complicated because they normally thrive within host tissues which exhibit properties that are difficult to replicate in vitro. This is especially true in the case of intracellular pathogens. However, careful replication of key features of the host environment can resolve these difficulties (e.g. host metabolites, dissolved oxygen), such as with the Q fever pathogen, Coxiella burnetii.

Viral Culture Methods

The culture of viruses requires the culture of cells of mammalian, plant, fungal or bacterial origin as hosts for the growth and replication of the virus. Whole wild type viruses, recombinant viruses or viral products may be generated in cell types other than their natural hosts under the right conditions. Depending on the species of the virus, infection and viral replication may result in host cell lysis and formation of a viral plaque.

Viral culture is a laboratory test in which samples are placed with a cell type that the virus being tested for is able to infect. If the cells show changes, known as cytopathic effects, then the culture is positive.

Traditional viral culture has been generally superseded by shell vial culture, in which the sample is centrifuged onto a single layer of cells and viral growth is measured by antigen detection methods. This greatly reduces the time to detection for slow growing viruses such as cytomegalovirus, for which the method was developed. In addition, the centrifugation step in shell vial culture enhances the sensitivity of this method because after centrifugation, the viral particles of the sample are in close proximity to the cells.

Human and monkey cells are used in both traditional viral culture and shell vial culture.

Human virus types that can be identified by viral culture include adenovirus, cytomegalovirus, enteroviruses, herpes simplex virus, influenza virus, parainfluenza virus, rhinovirus, respiratory syncytial virus, varicella zoster virus, measles and mumps. For these, the final identification method is generally by immunofluorescence, with exception of cytomegalovirus and rhinovirus, whose identification in a viral culture are determined by cytopathic effects.

Common Cell Lines

Human Cell Lines

- HeLa
- National Cancer Institute's 60 cancer cell lines
- ESTDAB database
- DU145 (prostate cancer)
- Lncap (prostate cancer)
- MCF-7 (breast cancer)

- MDA-MB-438 (breast cancer)
- PC3 (prostate cancer)
- T47D (breast cancer)
- THP-1 (acute myeloid leukemia)
- U87 (glioblastoma)
- SHSY5Y Human neuroblastoma cells, cloned from a myeloma
- Saos-2 cells (bone cancer).

Primate Cell Lines

- Vero (African green monkey Chlorocebus kidney epithelial cell line initiated in 1962).

Rat Tumor Cell Lines

- GH3 (pituitary tumor)
- PC12 (pheochromocytoma).

Mouse Cell Lines

- MC3T3 (embryonic calvarium).

Plant Cell Lines

- Tobacco BY-2 cells (kept as cell suspension culture, they are model system of plant cell).

Other Species Cell Lines

- Zebrafish ZF4 and AB9 cells
- Madin-Darby canine kidney (MDCK) epithelial cell line
- Xenopus A6 kidney epithelial cells.

List of Gene Prediction Software

In computational biology gene prediction or gene finding refers to the process of identifying the regions of genomic DNA that encode genes. This includes protein-coding genes as well as RNA genes, but may also include prediction of other functional elements such as regulatory regions. Gene finding is one of the first and most important steps in understanding the genome of a species once it has been sequenced.

In its earliest days, "gene finding" was based on painstaking experimentation on living cells and organisms. Statistical analysis of the rates of homologous recombination of several different genes could determine their order on a certain chromosome, and information from many such experiments could be combined to create a genetic map specifying the rough location of known genes relative to each other. Today, with comprehensive genome sequence and powerful computational resources at the disposal of the research community, gene finding has been redefined as a largely computational problem.

Determining that a sequence is functional should be distinguished from determining the function of the gene or its product. The latter still demands in vivo experimentation through gene knockout and other assays, although frontiers of bioinformatics research are making it increasingly possible to predict the function of a gene based on its sequence alone.

Extrinsic Approaches

In extrinsic (or evidence-based) gene finding systems, the target genome is searched for sequences that are similar to extrinsic evidence in the form of the known sequence of a messenger RNA (mRNA) or protein product. Given an mRNA sequence, it is trivial to derive a unique genomic DNA sequence from which it had to have been transcribed. Given a protein sequence, a family of possible coding DNA sequences can be derived by reverse translation of the genetic code. Once candidate DNA sequences have been determined, it is a relatively straightforward algorithmic problem to efficiently search a target genome for matches, complete or partial, and exact or inexact. BLAST is a widely used system designed for this purpose.

A high degree of similarity to a known messenger RNA or protein product is strong evidence that a region of a target genome is a protein-coding gene. However, to apply this approach systemically requires extensive sequencing of mRNA and protein products. Not only is this expensive, but in complex organisms, only a subset of all genes in the organism's genome are expressed at any given time, meaning that extrinsic evidence for many genes is not readily accessible in any single cell culture. Thus, in order to collect extrinsic evidence for most or all of the genes in a complex organism, many hundreds or thousands of different cell types must be studied, which itself presents further difficulties. For example, some human genes may be expressed only during development as an embryo or fetus , which might be difficult to study for ethical reasons.

Despite these difficulties, extensive transcript and protein sequence databases have been generated for human as well as other important model organisms in biology, such as mice and yeast. For example, the RefSeq database contains transcript and protein sequence from many different species, and the Ensembl system comprehensively maps this evidence to human and several other genomes. It is, however, likely that these databases are both incomplete and contain small but significant amounts of erroneous data.

Ab Initio Approaches

Because of the inherent expense and difficulty in obtaining extrinsic evidence for many genes, it is also necessary to resort to Ab initio gene finding, in which genomic DNA sequence alone is systematically searched for certain tell-tale signs of protein-coding genes. These signs can be broadly categorized as either signals, specific sequences that indicate the presence of a gene nearby, or content, statistical properties of protein-coding sequence itself. Ab initio gene finding might be more accurately characterized as gene prediction, since extrinsic evidence is generally required to conclusively establish that a putative gene is functional.

In the genomes of prokaryotes, genes have specific and relatively well-understood promoter sequences (signals), such as the Pribnow box and transcription factor binding sites, which are easy to systematically identify. Also, the sequence coding for a protein occurs as one contiguous open reading frame (ORF), which is typically many hundred or thousands of base pairs long. The statistics of stop codons are such that even finding an open reading frame of this length is a fairly informative sign. (Since 3 of the 64 possible codons in the genetic code are stop codons, one would expect a stop codon approximately every 20–25 codons, or 60–75 base pairs, in a random sequence.) Protein-coding DNA has certain periodicities and other statistical properties that are easy to detect in sequence of this length. These characteristics make prokaryotic gene finding relatively straightforward, and well-designed systems are able to achieve high levels of accuracy.

Ab initio gene finding in eukaryotes, especially complex organisms like humans, is considerably more challenging for several reasons. First, the promoter and other regulatory signals in these genomes are more complex and less well-understood than in prokaryotes, making them more difficult to reliably recognize. Two classic examples of signals identified by eukaryotic gene finders are CpG islands and binding sites for a poly(A) tail.

Second, splicing mechanisms employed by eukaryotic cells mean that a particular protein-coding sequence in the genome is divided into several parts (exons), separated by non-coding sequences (introns). (Splice sites are themselves another signal that eukaryotic gene finders are often designed to identify.) A typical protein-coding gene in humans might be divided into a dozen exons, each less than two hundred base pairs in length, and some as short as twenty to thirty. It is therefore much more difficult to detect periodicities and other known content properties of protein-coding DNA in eukaryotes.

Advanced gene finders for both prokaryotic and eukaryotic genomes typically use complex probabilistic models, such as hidden Markov models, in order to combine information from a variety of different signal and content measurements.

The GLIMMER system is a widely used and highly accurate gene finder for prokaryotes. GeneMark is another popular approach. Eukaryotic ab initio gene finders, by comparison, have achieved only limited success; notable examples are the GENSCAN and geneid programs. The SNAP gene finder is HMM-based like Genscan and attempts to be more adaptable to different organisms, addressing problems related to using a gene finder on a genome sequence that it was not trained against.

A few recent approaches like mSplicer, CONTRAST, or mGene also use machine learning techniques like support vector machines for successful gene prediction. They build a discriminative model using hidden Markov support vector machines or conditional random fields to learn an accurate gene prediction scoring function.

Other Signals

Among the derived signals used for prediction are statistics resulting from the sub-sequence statistics like k-mer statistics, Fourier transform of a pseudo-number-coded DNA, Z-curve parameters and certain run features. It has been suggested that signals other than those directly detectable in sequences may improve gene prediction.

For example, the role of secondary structure in the identification of regulatory motifs has been reported. In addition, it has been suggested that RNA secondary structure prediction helps splice site prediction.

Comparative Genomics Approaches

As the entire genomes of many different species are sequenced, a promising direction in current research on gene finding is a comparative genomics approach. This is based on the principle that the forces of natural selection cause genes and other functional elements to undergo mutation at a slower rate than the rest of the genome, since mutations in functional elements are more likely to negatively impact the organism than mutations elsewhere. Genes can thus be detected by comparing the genomes of related species to detect this evolutionary pressure for conservation. This approach was first applied to the mouse and human genomes, using programs such as SLAM, SGP and Twinscan/N-SCAN.

Comparative gene finding can also be used to project high quality annotations from one genome to another. Notable examples include Projector, GeneWise and GeneMapper. Such techniques now play a central role in the annotation of all genomes.

Notable Genes

This list describes some of the most notable genes present in the human genome.

Note: this does not attempt to be an exhausive list of genes, of which there are tens of thousands for humans alone: this is intended to be a list of genes which are particularly notable in scientific research or in their role in relation to disease, or have featured in a recent significant research paper or news story.

Genes Notable for their Function

Meme

A meme is "an idea, behaviour or style that spreads from person to person within a culture."A meme acts as a unit for carrying cultural ideas, symbols or practices, which can be transmitted from one mind to another through writing, speech, gestures, rituals or other imitable phenomena.

Supporters of the concept regard memes as cultural analogues to genes in that they self-replicate, mutate and respond to selective pressures.

The word 'meme' is a shortening (modeled on 'gene') of 'mimeme' (from ancient Greek "something imitated", from "to imitate", from mimos "mime") and it was coined by the British evolutionary biologist Richard Dawkins in *The Selfish Gene* (1976) as a concept for discussion of evolutionary principles in explaining the spread of ideas and cultural phenomena.

Examples of memes given in the book included melodies, catch-phrases, fashion and the technology of building arches.

Advocates of the meme idea say that memes may evolve by natural selection in a manner analogous to that of biological evolution. Memes do this through the processes of variation, mutation, competition and inheritance, each of which influence a meme's reproductive success.

Memes spread through the behaviours that they generate in their hosts. Memes that propagate less prolifically may become extinct, while others may survive, spread and (for better or for worse) mutate. Memes that replicate most effectively enjoy more success. Some memes may replicate effectively even when they prove to be detrimental to the welfare of their hosts.

A field of study called memetics arose in the 1990s to explore the concepts and transmission of memes in terms of an evolutionary model. Criticism from a variety of fronts has challenged the notion that scholarship can examine memes empirically. Developments in neuroimaging may however make empirical study possible. Some commentators question the idea that one can meaningfully categorize culture in terms of discrete units.

Origins

Max Stirner's (1844) *The Ego and Its Own* puts forth the idea that the individual is dominated by illusory concepts ('fixed ideas' or 'spooks'), which can be shaken and undermined by each individual, though he does not use the term meme for this. He offers examples such as nationalism and religion.

Historically, the notion of a unit of social evolution, and a similar term (from Greek mneme, "memory"), first appeared in 1904 in a work by the German Lamarckist biologist Richard Semon titled Die Mnemischen Empfindungen in ihren Beziehungen zu den Originalempfindungen (loosely translatable as "Memory-feelings in relation to original feelings"). According to the OED, the word mneme appears in English in 1921 in L. Simon's translation of Semon's book: *The Mneme*. Bertrand Russell in *The Analysis of Mind* (1921) summarizes Semon's work and furthers its application to the analysis of memory. Discussion of mnemes plays a central role in Russell's book, but he does not explicitly make comparisons to the gene concept.

Laurent noted the use of the term mneme in Maurice Maeterlinck's *The Life of the White Ant* (1926), and has highlighted similarities to Dawkins' concept.

The analogy between culturally transmitted information and genetically transmitted information was perceived clearly enough by Luigi Luca Cavalli-Sforza and Marcus W. Feldman to allow them to formulate and analyze quantitative models of cultural transmission and selection. They published a series of papers beginning in 1973.

The word meme originated with Dawkins' (1976) book *The Selfish Gene*. To emphasize commonality with genes, Dawkins coined the term "meme" by shortening "mimeme", which derives from the Greek word mimema ("something imitated").

Dawkins states that he did not know of the "mneme", and said that he wanted "a monosyllable that sounds a bit like 'gene'". Dawkins wrote that evolution depended not on the particular chemical basis of genetics, but only on the existence of a self-replicating unit of transmission – in the case of biological evolution, the gene. For Dawkins, the meme exemplified another self-replicating unit with potential significance in explaining human behavior and cultural evolution.

Concept

Dawkins used the term to refer to any cultural entity that an observer might consider a replicator. He hypothesised that one could view many cultural entities as replicators, and pointed to melodies, fashions and learned skills as examples. Memes generally replicate through exposure to humans, who have evolved as efficient copiers of information and behaviour. Because humans do not always copy memes perfectly, and because they may refine, combine or otherwise modify them with other memes to create new memes, they can change over time. Dawkins likened the process by which memes survive and change through the evolution of culture to the natural selection of genes in biological evolution.

Dawkins defined the meme as a unit of cultural transmission, or a unit of imitation and replication, but later definitions would vary. Memes, analogously to genes, vary in their aptitude to replicate; memes which are good at getting themselves copied tend to spread and remain, whereas the less good ones have a higher probability of being ignored and forgotten. Thus "better" memes are selected. The lack of a consistent, rigorous, and precise understanding of what typically makes up one unit of cultural transmission remains a problem in debates about memetics. In contrast, the concept of genetics gained concrete evidence with the discovery of the biological functions of DNA. Meme transmission does not necessarily require a physical medium, unlike genetic transmission.

Transmission

Life-forms can transmit information both vertically (from parent to child, via replication of genes) and horizontally (through viruses and other means). Malcolm Gladwell wrote, "A meme is an idea that behaves like a virus—that moves through a population, taking hold in each person it infects." Memes can replicate vertically or horizontally within a single biological generation. They may also lie dormant for long periods of time. Memes spread by the behaviors that they generate in their hosts. Imitation counts as an important characteristic in the propagation of memes. Imitation often involves the copying of an observed behaviour of another individual, but memes may transmit from one individual to another through a copy recorded in an inanimate source, such as a book or a musical score. McNamara has suggested that memes can be thereby classified as either internal or external memes, (i-memes or e-memes). Researchers have observed memetic copying in just a few species on Earth, including hominids, dolphins and birds (that learn how to sing by imitating their parents or neighbours).

Some commentators have likened the transmission of memes to the spread of contagions. Social contagions such as fads, hysteria, copycat crime, and copycat suicide exemplify memes seen as the contagious imitation of ideas. Observers distinguish the contagious imitation of memes from instinctively contagious phenomena such as yawning and laughing, which they consider innate (rather than socially learned) behaviors.

Aaron Lynch described seven general patterns of meme transmission, or "thought contagion":

1. Quantity of parenthood: an idea that influences the number of children one has. Children respond particularly receptively to the ideas of their parents, and thus ideas that directly or indirectly encourage a higher birthrate will replicate themselves at a higher rate than those that discourage higher birthrates.

2. Efficiency of parenthood: an idea that increases the proportion of children who will adopt ideas of their parents. Cultural separatism exemplifies one practice in which one can expect a higher rate of meme-replication — because the meme for separation creates a barrier from exposure to competing ideas.
3. Proselytic: ideas generally passed to others beyond one's own children. Ideas that encourage the proselytism of a meme, as seen in many religious or political movements, can replicate memes horizontally through a given generation, spreading more rapidly than parent-to-child meme-transmissions do.
4. Preservational: ideas that influence those that hold them to continue to hold them for a long time. Ideas that encourage longevity in their hosts, or leave their hosts particularly resistant to abandoning or replacing these ideas, enhance the preservability of memes and afford protection from the competition or proselytism of other memes.
5. Adversative: ideas that influence those that hold them to attack or sabotage competing ideas and/or those that hold them. Adversative replication can give an advantage in meme transmission when the meme itself encourages aggression against other memes.
6. Cognitive: ideas perceived as cogent by most in the population who encounter them. Cognitively transmitted memes depend heavily on a cluster of other ideas and cognitive traits already widely held in the population, and thus usually spread more passively than other forms of meme transmission. Memes spread in cognitive transmission do not count as self-replicating.
7. Motivational: ideas that people adopt because they perceive some self-interest in adopting them. Strictly speaking, motivationally transmitted memes do not self-propagate, but this mode of transmission often occurs in association with memes self-replicated in the efficiency parental, proselytic and preservational modes.

Memes as Discrete Units

Richard Dawkins initially defined meme as a noun that "conveys the idea of a unit of cultural transmission, or a unit of imitation". John S. Wilkins retained the notion of meme as a kernel of cultural imitation while emphasizing the meme's evolutionary aspect, defining the meme as "the least unit of sociocultural information relative to a selection process that has favourable or unfavourable selection bias that exceeds its endogenous tendency to change." The meme as a unit provides a convenient means of discussing "a piece of thought copied from person to person", regardless if that thought contains others inside it, or forms part of a larger meme. A meme could consist of a single word, or a meme could consist of the entire speech in which that word first occurred. This forms an analogy to the idea of a gene as a single unit of self-replicating information found on the self-replicating chromosome.

While the identification of memes as "units" conveys their nature to replicate as discrete, indivisible entities, it does not imply that thoughts somehow become quantized or that "atomic" ideas exist that cannot be dissected into smaller pieces. A meme has no given size. Susan Blackmore writes that melodies from Beethoven's symphonies are commonly used to illustrate the difficulty involved in delimiting memes as discrete units. She notes that while the first four notes of Beethoven's Fifth Symphony form a meme widely replicated as an independent unit, one can regard the entire symphony as a single meme as well. The inability to pin an idea or cultural feature to quantifiable key units is widely acknowledge as a problem for memetics. It has been argued however that the traces of memetic processing can be quantified utilizing neuroimaging techniques which measure changes in the connectivity profiles between brain regions." Blackmore meets such criticism by stating that memes compare with genes in this respect: that while a gene has no particular size, nor can we ascribe every phenotypic feature directly to a particular gene, it has value because it encapsulates that key unit of inherited expression subject to evolutionary pressures. To illustrate, she notes evolution selects for the gene for features such as eye color; it does not select for the individual nucleotide in a strand of DNA. Memes play a comparable role in understanding the evolution of imitated behaviors.

The 1981 book *Genes, Mind, and Culture: The Coevolutionary Process* by Charles J. Lumsden and E. O. Wilson proposed the theory that genes and culture co-evolve, and that the fundamental biological units of culture must correspond to neuronal networks that function as nodes of semantic memory. They coined their own term, "culturgen", which did not catch on. Coauthor Wilson later acknowledged the term meme as the best label for the fundamental unit of cultural inheritance in his 1998 book *Consilience: The Unity of Knowledge*, which elaborates upon the fundamental role of memes in unifying the natural and social sciences.

Evolutionary Influences on Memes

Richard Dawkins noted the three conditions that must exist for evolution to occur:

1. variation, or the introduction of new change to existing elements;
2. heredity or replication, or the capacity to create copies of elements;
3. differential "fitness", or the opportunity for one element to be more or less suited to the environment than another.

Dawkins emphasizes that the process of evolution naturally occurs whenever these conditions co-exist, and that evolution does not apply only to organic elements such as genes.

He regards memes as also having the properties necessary for evolution, and thus sees meme evolution as not simply analogous to genetic evolution, but as a real phenomenon subject to the laws of natural selection. Dawkins noted that as various ideas pass from one generation to the next, they may either enhance or detract from the survival of the people who obtain those ideas, or influence the survival of the ideas themselves. For example, a certain culture may develop unique designs and methods of tool-making that give it a competitive advantage over another culture. Each tool-design thus acts somewhat similarly to a biological gene in that some populations have it and others do not, and the meme's function directly affects the presence of the design in future generations. In keeping with the thesis that in evolution one can regard organisms simply as suitable "hosts" for reproducing genes, Dawkins argues that one can view people as "hosts" for replicating memes. Consequently, a successful meme may or may not need to provide any benefit to its host.

Unlike genetic evolution, memetic evolution can show both Darwinian and Lamarckian traits. Cultural memes will have the characteristic of Lamarckian inheritance when a host aspires to replicate the given meme through inference rather than by exactly copying it. Take for example the case of the transmission of a simple skill such as hammering a nail, a skill that a learner imitates from watching a demonstration without necessarily imitating every discrete movement modeled by the teacher in the demonstration, stroke for stroke. Susan Blackmore distinguishes the difference between the two modes of inheritance in the evolution of memes, characterizing the Darwinian mode as "copying the instructions" and the Lamarckian as "copying the product."

Clusters of memes, or memeplexes (also known as meme complexes or as memecomplexes), such as cultural or political doctrines and systems, may also play a part in the acceptance of new memes. Memeplexes comprise groups of memes that replicate together and coadapt. Memes that fit within a successful memeplex may gain acceptance by "piggybacking" on the success of the memeplex. As an example, John D.

Gottsch discusses the transmission, mutation and selection of religious memeplexes and the theistic memes contained. Theistic memes discussed include the "prohibition of aberrant sexual practices such as incest, adultery, homosexuality, bestiality, castration, and religious prostitution", which may have increased vertical transmission of the parent religious memeplex. Similar memes are thereby included in the majority of religious memeplexes, and harden over time; they become an "inviolable canon" or set of dogmas, eventually finding their way into secular law. This could also be referred to as the propagation of a taboo.

Memetics

The discipline of memetics, which dates from the mid 1980s, provides an approach to evolutionary models of cultural information transfer based on the concept of the meme. Memeticists have proposed that just as memes function analogously to genes, memetics functions analogously to genetics. Memetics attempts to apply conventional scientific methods (such as those used in population genetics and epidemiology) to explain existing patterns and transmission of cultural ideas.

Principal criticisms of memetics include the claim that memetics ignores established advances in other fields of cultural study, such as sociology, cultural anthropology, cognitive psychology, and social psychology. Questions remain whether or not the meme concept counts as a validly disprovable scientific theory. This view regards memetics as a theory in its infancy: a protoscience to proponents, or a pseudoscience to some detractors.

Criticism of Meme Theory

An objection to the study of the evolution of memes in genetic terms (although not to the existence of memes) involves a perceived gap in the gene/ meme analogy: the cumulative evolution of genes depends on biological selection-pressures neither too great nor too small in relation to mutation-rates. There seems no reason to think that the same balance will exist in the selection pressures on memes.

Luis Benitez-Bribiesca M.D., a critic of memetics, calls the theory a "pseudoscientific dogma" and "a dangerous idea that poses a threat to the serious study of consciousness and cultural evolution". As a factual criticism, Benitez-Bribiesca points to the lack of a "code script" for memes (analogous to the DNA of genes), and to the excessive instability of the meme mutation mechanism (that of an idea going from one brain to another), which would lead to a low replication accuracy and a high mutation rate, rendering the evolutionary process chaotic. British political philosopher John Gray has characterized Dawkins' memetic theory of religion as "nonsense" and "not even a theory... the latest in a succession of ill-judged Darwinian metaphors", comparable to Intelligent Design in its value as a science.

Another critique comes from semiotic theorists such as Deacon and Kull This view regards the concept of "meme" as a primitivized concept of "sign". The meme is thus described in memetics as a sign lacking a triadic nature. Semioticians can regard a meme as a "degenerate" sign, which includes only its ability of being copied. Accordingly, in the broadest sense, the objects of copying are memes, whereas the objects of translation and interpretation are signs. Fracchia and Lewontin regard memetics as reductionist and inadequate.

Potential Lack of Philosophical Depth

In his chapter titled "Truth" published in the Encyclopedia of *Phenomenology*, Dieter Lohmar questions the memeticists' reduction of the highly complex body of ideas (such as religion, politics, war, justice, and science itself) to a putatively one-dimensional series of memes. He sees memes as an abstraction and such a reduction as failing to produce greater understanding of those ideas. The highly interconnected, multi-layering of ideas resists memetic simplification to an atomic or molecular form; as does the fact that each of our lives remains fully enmeshed and involved in such "memes". Lohmar argues that one cannot view memes through a microscope in the way one can detect genes. The leveling-off of all such interesting "memes" down to some neutralized molecular "substance" such as "meme-substance" introduces a bias toward "scientism" and abandons the very essence of what makes ideas interesting, richly available, and worth studying.

Applications

Opinions differ as to how best to apply the concept of memes within a "proper" disciplinary framework. One view sees memes as providing a useful philosophical perspective with which to examine cultural evolution. Proponents of this view (such as Susan Blackmore and Daniel Dennett) argue that considering cultural developments from a meme's-eye view—as if memes themselves respond to pressure to maximise their own replication and survival—can lead to useful insights and yield valuable predictions into how culture develops over time. Others such as Bruce Edmonds and Robert Aunger have focused on the need to provide an empirical grounding for memetics to become a useful and respected scientific discipline. A third approach, described as "radical memetics", seeks to place memes at the centre of a materialistic theory of mind and of personal identity.

Prominent researchers in evolutionary psychology and anthropology, including Scott Atran, Dan Sperber, Pascal Boyer, John Tooby and others, argue the possibility of incompatibility between modularity of mind and memetics. In their view, minds structure certain communicable aspects of the ideas produced, and these communicable aspects generally trigger or elicit ideas in other minds through inference (to relatively rich structures generated from often low-fidelity input) and not high-fidelity replication or imitation. Atran discusses communication involving religious beliefs as a case in point. In one set of experiments he asked religious people to write down on a piece of paper the meanings of the Ten Commandments. Despite the subjects' own expectations of consensus, interpretations of the commandments showed wide ranges of variation, with little evidence of consensus. In another experiment, subjects with autism and subjects without autism interpreted ideological and religious sayings (for example, "Let a thousand flowers bloom" or "To everything there is a season"). People with autism showed a significant tendency to closely paraphrase and repeat content from the original statement (for example: "Don't cut flowers before they bloom"). Controls tended to infer a wider range of cultural meanings with little replicated content (for example: "Go with the flow" or "Everyone should have equal opportunity"). Only the subjects with autism—who lack the degree of inferential capacity normally associated with aspects of theory of mind—came close to functioning as "meme machines".

In his book *The Robot's Rebellion*, Stanovich uses the memes and memeplex concepts to describe a program of cognitive reform that he refers to as a "rebellion". Specifically, Stanovich argues that the use of memes as a descriptor for cultural units is beneficial because it serves to emphasize transmission and acquisition properties that parallel the study of epidemiology. These properties make salient the sometimes parasitic nature of acquired memes, and as a result individuals should be motivated to reflectively acquire memes using what he calls a "Neurathian bootstrap" process.

Religion

Although social scientists such as Max Weber sought to understand and explain religion in terms of a cultural attribute, Richard Dawkins called for a re-analysis of religion in terms of the evolution of self-replicating ideas apart from any resulting biological advantages they might bestow. As an enthusiastic Darwinian, I have been dissatisfied with explanations that my fellow-enthusiasts have offered for human behaviour. They have tried to look for 'biological advantages' in various attributes of human civilization. For instance, tribal religion has been seen as a mechanism for solidifying group identity, valuable for a pack-hunting species whose individuals rely on cooperation to catch large and fast prey. Frequently the evolutionary preconception in terms of which such theories are framed is implicitly group-selectionist, but it is possible to rephrase the theories in terms of orthodox gene selection.

—Richard Dawkins, *The Selfish Gene*

He argued that the role of key replicator in cultural evolution belongs not to genes, but to memes replicating thought from person to person by means of imitation. These replicators respond to selective pressures that may or may not affect biological reproduction or survival.

In her book *The Meme Machine*, Susan Blackmore regards religions as particularly tenacious memes. Many of the features common to the most widely practiced religions provide built-in advantages in an evolutionary context, she writes. For example, religions that preach of the value of faith over evidence from everyday experience or reason inoculate societies against many of the most basic tools people commonly use to evaluate their ideas. By linking altruism with religious affiliation, religious memes can proliferate more quickly because people perceive that they can reap societal as well as personal rewards.

The longevity of religious memes improves with their documentation in revered religious texts. Aaron Lynch attributed the robustness of religious memes in human culture to the fact that such memes incorporate multiple modes of meme transmission. Religious memes pass down the generations from parent to child and across a single generation through the meme-exchange of proselytism. Most people will hold the religion taught them by their parents throughout their life.

Many religions feature adversarial elements, punishing apostasy, for instance, or demonizing infidels. In *Thought Contagion Lynch* identifies the memes of transmission in Christianity as especially powerful in scope. Believers view the conversion of non-believers both as a religious duty and as an act of altruism. The promise of heaven to believers and threat of hell to non-believers provide a strong incentive for members to retain their belief.

Lynch asserts that belief in the Crucifixion of Jesus in Christianity amplifies each of its other replication advantages through the indebtedness believers have to their Savior for sacrifice on the cross. The image of the crucifixion recurs in religious sacraments, and the proliferation of symbols of the cross in homes and churches potently reinforces the wide array of Christian memes.

Although religious memes have proliferated in human cultures, the modern scientific community has been relatively resistant to religious belief. Robertson (2007) reasoned that if evolution is accelerated in conditions of propagative difficulty , then we would expect to encounter variations of religious memes, established in general populations, addressed to scientific communities. Using a memetic approach, Robertson deconstructed two attempts to privilege religiously held spirituality in scientific discourse. Advantages of a memetic approach as compared to more traditional "modernization" and "supply side" theses in understanding the evolution and propagation of religion were explored.

Memetic Explanations of Racism

In Cultural Software: A Theory of Ideology, Jack Balkin argued that memetic processes can explain many of the most familiar features of ideological thought. His theory of "cultural software" maintained that memes form narratives, networks of cultural associations, metaphoric and metonymic models, and a variety of different mental structures. Balkin maintains that the same structures used to generate ideas about free speech or free markets also serve to generate racist beliefs.

To Balkin, whether memes become harmful or maladaptive depends on the environmental context in which they exist rather than in any special source or manner to their origination. Balkin describes racist beliefs as "fantasy" memes that become harmful or unjust "ideologies" when diverse peoples come together, as through trade or competition.

Internet Culture

The term "Internet meme" refers to a catchphrase or concept that spreads rapidly from person to person via the Internet, largely through Internet-based email, blogs, forums, Imageboards, social networking sites and instant messaging.

Meme Maps

One technique of meme mapping represents the evolution and transmission of a meme across time and space. Diagram (an analemma) to map the gestation (in the lower loop), birth (at the choke point), and development (in the upper loop) of the selected meme. Such meme maps are non-scalar, with time mapped onto the y-axis and space onto the x-axis transect. One can read the temporal progress of the mapped meme from south to north on such a meme map. Paull has published a worked example using the "organics meme" (as in organic agriculture).

Robertson (2010) used a second technique of meme mapping to create two-dimensional representations of the selves of eleven participants drawn from both individualist and collectivist cultures. Participant narratives were transcribed, segmented and coded using a method similar to grounded theory. Coded segments exhibiting referent, connotative, affective and behavioral dimensions were declared to be memes. Memes that shared connotative, affective or behavioural qualities were linked. All of the maps in Robertson's sample evidenced volition, constancy, uniqueness, production, intimacy, and social interest. This method of mapping the self was successfully used in therapy to treat a youth who had attempted suicide on five occasions (Robertson, 2011). The youth and psychotherapist co-constructed a plan to change the youth's presenting self, and her progress in making those changes was tracked in subsequent self-maps.

Dual Inheritance Theory

Dual inheritance theory (DIT), also known as gene-culture coevolution, was developed in the late 1970s and early 1980s to explain how human behavior is a product of two different and interacting evolutionary processes: genetic evolution and cultural evolution. DIT is a "middle-ground" between much of social science , which views culture as the primary cause of human behavioral variation, and human sociobiology and evolutionary psychology which view culture as a by-product of genetic selection.

In DIT, culture is defined as information in human brains that got there by social learning. Cultural evolution is considered a Darwinian selection process that acts on cultural information. Dual inheritance theorists often describe this by analogy to genetic evolution, which is a Darwinian selection process acting on genetic information. Because genetic evolution is relatively well understood, most of DIT examines cultural evolution and the interactions between cultural evolution and genetic evolution.

Theoretical Basis

DIT holds that genetic and cultural evolution interact in the evolution of Homo sapiens. DIT recognizes that the natural selection of genotypes is an important component of the evolution of human behavior and that cultural traits can be constrained by genetic imperatives. However, DIT also recognizes that genetic evolution has endowed the human species with a parallel evolutionary process of cultural evolution. DIT makes three main claims:

Culture Capacities are Adaptations

The human capacity to store and transmit culture arose from genetically evolved psychological mechanisms. This implies that at some point during the evolution of the human species a type of social learning leading to cumulative cultural evolution was evolutionarily advantageous.

Culture Evolves

Social learning processes give rise to cultural evolution. Cultural traits are transmitted differently than genetic traits and, therefore, result in different population-level effects. These effects can help explain human behavioural variation.

Genes and Culture Coevolve

Cultural traits alter the social and physical environments under which genetic selection operates. For example, the cultural adoptions of agriculture and dairying have, in humans, caused genetic selection for the traits to digest starch and lactose, respectively. As another example, it is likely that once culture became adaptive, genetic selection caused a refinement of the cognitive architecture that stores and transmits cultural information. This refinement may have further influenced the way culture is stored and the biases that govern its transmission.

DIT also predicts that, under certain situations, cultural evolution may select for traits that are genetically maladaptive. An example of this is the demographic transition, which describes the fall of birth rates within industrialized societies. Dual inheritance theorists hypothesize that the demographic transition may be a result of a prestige bias, where individuals that forgo reproduction to gain more influence in industrial societies are more likely to be chosen as cultural models.

View of Culture

People have defined the word "culture" to describe a large set of different phenomena. A definition that sums up what is meant by "culture" in DIT is:

Culture is information stored in individuals' brains that is capable of affecting behavior and that got there through social learning. This view of culture emphasizes population thinking by focusing on the process by which culture is generated and maintained. It also views culture as a dynamic property of individuals, as opposed to a view of culture as a superorganic entity to which individuals must conform. This view's main advantage is that it connects individual-level processes to population-level outcomes.

Genetic Influence on Cultural Evolution

Genes have an impact on cultural evolution via psychological predispositions on cultural learning. Genes encode much of the information needed to form the human brain. Genes constrain the brain's structure and, hence, the ability of the brain to acquire and store culture. Genes may also endow individuals with certain types of transmission bias (described below).

Cultural Influences on Genetic Evolution

Culture can profoundly influence gene frequencies in a population. One of the best known examples is the prevalence of the genotype for adult lactose absorption in human populations, such as Northern Europeans and some African societies, with a long history of raising cattle for milk. Other societies such as East Asians and Amerindians, retain the typical mammalian genotype in which the body shuts down lactase production shortly after the normal age of weaning. This implies that the cultural practice of raising cattle for milk led to a selection for genetic traits for lactose digestion. Recently, analysis of natural selection on the human genome suggests that civilization has accelerated genetic change in humans over the past 10,000 years.

Cumulative Cultural Evolution

In DIT, the evolution of culture is dependent on the evolution of social learning. Analytic models show that social learning becomes evolutionarily beneficial when the environment changes with enough frequency that genetic inheritance can not track the changes, but not fast enough that individual learning is more efficient. While other species have social learning, and thus some level of culture, only humans, some birds and chimpanzees are known to have cumulative culture. Boyd and Richerson argue that the evolution of cumulative culture depends on observational learning and is uncommon in other species because it is ineffective when it is rare in a population. They propose that the environmental changes occurring in the Pleistocene may have provided the right environmental conditions. Michael Tomasello argues that cumulative cultural evolution results from a "ratchet effect" that began when humans developed the cognitive architecture to understand others as mental agents. Tomasello proposed in the 80s that there are some disparities between the observational learning mechanisms found in humans and great apes -

which go some way to explain the observable difference between great ape traditions and human types of culture.

Cultural Group Selection

Although group selection is commonly thought to be nonexistent or unimportant in genetic evolution, DIT predicts that, due to the nature of cultural inheritance, it may be an important force in cultural evolution. The reason group selection is thought to operate in cultural evolution is because of conformist biases. Conformist biases make it difficult for novel cultural traits to spread through a population. Conformist bias also helps maintain variation between groups. These two properties, rare in genetic transmission, are necessary for group selection to operate. Based on an earlier model by Cavalli-Sforza and Feldman, Boyd and Richerson show that conformist biases are almost inevitable when traits spread through social learning, implying that group selection is common in cultural evolution. Analysis of small groups in New Guinea imply that cultural group selection might be a good explanation for slowly changing aspects of social structure, but not for rapidly changing fads. The ability of cultural evolution to maintain intergroup diversity is what allows for the study of cultural phylogenetics.

Historical Development

The idea that human cultures undergo a similar evolutionary process as genetic evolution goes back at least to Darwin. In the 1960s, Donald T. Campbell published some of the first theoretical work that adapted principles of evolutionary theory to the evolution of cultures. In 1976, two developments in cultural evolutionary theory set the stage for DIT.

In that year Richard Dawkins's *The Selfish Gene* introduced ideas of cultural evolution to a popular audience. Although one of the best-selling science books of all time, because of its lack of mathematical rigor, it had little impact on the development of DIT. Also in 1976, geneticists Marcus Feldman and Luigi Luca Cavalli-Sforza published the first dynamic models of gene-culture coevolution. These models were to form the basis for subsequent work on DIT, heralded by the publication of three seminal books in 1980 and 1981.

The first was Charles Lumsden and E.O. Wilson's *Genes, Mind and Culture*. This book outlined a series of mathematical models of how genetic evolution might favor the selection of cultural traits and how cultural traits might, in turn, affect the speed of genetic evolution. While it was the first book published describing how genes and culture might coevolve, it had relatively little impact on the further development of DIT. Some critics felt that their models depended too heavily on genetic mechanisms at the expense of cultural mechanisms. Controversy surrounding Wilson's sociobiological theories may also have decreased the lasting impact of this book.

The second book (1981) was Cavalli-Sforza and Feldman's *Cultural Transmission and Evolution: A Quantitative Approach*. Borrowing heavily from population genetics and epidemiology, this book built a mathematical theory concerning the spread of cultural traits.

It describes the evolutionary implications of vertical transmission, passing cultural traits from parents to offspring; oblique transmission, passing cultural traits from any member of an older generation to a younger generation; and horizontal transmission, passing traits between members of the same population. The next significant DIT publication was Robert Boyd and Peter Richerson's (1985) *Culture and the Evolutionary Process*. This book presents the now-standard mathematical models the evolution of social learning under different environmental conditions, the population effects of social learning, various forces of selection on cultural learning rules, different forms of biased transmission and their population-level effects, and conflicts between cultural and genetic evolution. The book's conclusion also outlined areas for future research that are still relevant today.

Current and Future Research

In their book, Boyd and Richerson (1985) outlined an agenda for future DIT research. This agenda, outlined below, called for the development of both theoretical models and empirical research. DIT has since built a rich tradition of theoretical models over the past two decades. However, there has not been a comparable level of empirical work.

In a 2006 interview Harvard biologist E. O. Wilson expressed disappointment at the little attention afforded to DIT:

"...for some reason I haven't fully fathomed, this most promising frontier of scientific research has attracted very few people and very little effort."

Kevin Laland and Gillian Brown attribute this lack of attention to DIT's heavy reliance on formal modeling.

"In many ways the most complex and potentially rewarding of all approaches, [DIT], with its multiple processes and cerebral onslaught of sigmas and deltas, may appear too abstract to all but the most enthusiastic reader. Until such a time as the theoretical hieroglyphics can be translated into a respectable empirical science most observers will remain immune to its message."

Economist Herbert Gintis disagrees with this critique, citing empirical work as well as more recent work using techniques from behavioral economics. These behavioral economic techniques have been adapted to test predictions of cultural evolutionary models in laboratory settings as well as studying differences in cooperation in fifteen small-scale societies in the field.

Since one of the goals of DIT is to explain the distribution of human cultural traits, ethnographic and ethnologic techniques may also be useful for testing hypothesis stemming from DIT. Although findings from traditional ethnologic studies have been used to buttress DIT arguments, thus far there have been little ethnographic fieldwork designed to explicitly test these hypotheses. A major difficulty in using existing ethnographic data to test DIT is that traditional ethnographic data tends to ignore individual and intragroup cultural variation and focus almost entirely on intergroup variation.

DIT has been viewed as having great potential for unifying diverse academic fields under one overarching theory. Mesoudi, et al. have identified DIT as the ideal way to build a comprehensive theory of cultural evolution to answer questions about human behavior at different temporal and spacial scales.

Along with game theory, Herb Gintis has named DIT one of the two major conceptual theories with potential for unifying the behavioural sciences, including economics, biology, anthropology, sociology, psychology and political science. Because it addresses both the genetic and cultural components of human inheritance, Gintis sees DIT models as providing the best explanations for the ultimate cause of human behaviour and the best paradigm for integrating those disciplines with evolutionary theory. In a review of competing evolutionary perspectives on human behavior, Laland and Brown see DIT as the best candidate for uniting the other evolutionary perspectives under one theoretical umbrella.

Relation to Other Fields

Sociology and Cultural Anthropology

Two major topics of study in both sociology and cultural anthropology are human cultures and cultural variation. However, Dual inheritance theorists charge that both disciplines too often treat culture as a static superorganic entity that dictates human behaviour. Cultures are defined by a suite of common traits shared by a large group of people. DIT theoriests argue that this does not sufficiently explain variation in cultural traits at the individual level. By contrast, DIT models human culture at the individual level and views culture as the result of a dynamic evolutionary process at the population level.

Human Sociobiology and Evolutionary Psychology

Human sociobiologists and evolutionary psychologists try to understand how maximizing genetic fitness, in either the modern era or past environments, can explain human behaviour. To the human sociobiologist and evolutionary psychologist, culture is either trivial or so bound by genetic fitness that it is unimportant. When faced with a common and seemingly maladaptive trait, practitioners from these disciplines try to determine how the trait actually increases genetic fitness (maybe through kin selection or by speculating about early evolutionary environments). Dual Inheritance theorists, in contrast, will consider a variety of genetic and cultural processes in addition to natural selection on genes.

Human Behavioral Ecology

Human behavioural ecology (HBE) and DIT have a similar relationship to what ecology and evolutionary biology have in the biological sciences. HBE is more concerned about ecological process and DIT more focused on historical process. One difference is that human behavioural ecologists often assume that culture is a system that produces the most adaptive outcome in a given environment. This implies that similar behavioural traditions should be found in similar environments. However, this is not always the case. A study of African cultures showed that cultural history was a better predictor of cultural traits than local ecological conditions.

Memetics

Memetics, which comes from the meme idea described in Dawkins's *The Selfish Gene*, is similar to DIT in that it treats culture as an evolutionary process that is distinct from genetic transmission. However, there are some philosophical differences between memetics and DIT. One difference is that memetics' focus is on the selection potential of discrete replicators (memes), where DIT allows for transmission of both non-replicators and non-discrete cultural variants. DIT does not assume that replicators are necessary for cumulative adaptive evolution. DIT also more strongly emphasizes the role of genetic inheritance in shaping the capacity for cultural evolution. But perhaps the biggest difference is a difference in academic lineage. Memetics as a label is more influential in popular culture than in academia. Critics of memetics argue that it is lacking in empirical support or is conceptually ill-founded, and question whether there is hope for the memetic research program succeeding. Proponents point out that many cultural traits are discrete, and that many existing models of cultural inheritance assume discrete cultural units, and hence involve memes.

Memetic Engineering

Memetic engineering is a term developed and coined by Leveious Rolando, John Sokol, and Gibran Burchett while they researched and observed the behaviour of people after being purposely exposed (knowingly and unknowingly) to certain memetic themes. The term is based on Richard Dawkins' theory of memes.

- The process of developing memes, through meme-splicing and memetic synthesis, with the intent of altering the behaviour of others in society or humanity.

- The process of creating and developing theories or ideologies based on an analytical study of societies, cultures, their ways of thinking and the evolution of their minds.
- The process of modifying human beliefs, thought patterns, etc.

Definition

According to the theory , the effect a meme has on society is based on the application of the meme after understanding the qualities essential to the meme. For example, Rolando, Burchett and Sokol expand on their concept and explain that "Race" and "Racism" are memes incorporating several other memes, some of which have positive connotations in societies that reject racism.

According to the theory Memetic Engineering is, simply put, the analysis of an individual or individual's behavior, the selection of specific memes and the distribution or propagation of those memes with the intent of altering the behavior of others. A memetic engineer doesn't particularly have to consciously make the decision to alter another individuals behavior. It can happen unconsciously when specific behaviour is observed, transmitted and then replicated within the observer.

The process of creating and developing theories or ideologies based on an analytical study of societies, cultures, their ways of thinking and the evolution of their minds. Memes themselves are neither good nor bad. For example Race is an ideology that is made up of several memes. When a Meme is introduced, those concepts begin to take on their own process of evolution based on the person who adopts the ideology, internalizes it, and reintroduces it into society causing it to spread like a virus.

According to the above theory, typical memetic engineers include scientists, engineers, industrial designers, ad-men, artists, publicists, political activists, and religious missionaries. Dawkins agrees that much of theology and other theoretical aspects of religion can be viewed as the careful, even worshipful, handling of extremely powerful memeplexes with very odd or difficult traits.

Origins of Memetic Engineering

Memetic Engineering developed from diverse influences, including cutting-edge physics of consciousness and memetics research, chaos theory, semiotics, culture jamming, military information warfare, and the viral texts of iconoclasts William S. Burroughs, J.G. Ballard, and Genesis P-Orridge. It draws upon Third Culture sciences and conceptual worldviews for Social Engineering, Values Systems Alignment, and Culture Jamming purposes. An important example of macro-historical memetic engineering analysis explaining how domination, patriarchy, war and violence are culturally programmed is Riane Eisler's *The Chalice and the Blade* (San Francisco: Harper SanFrancisco,

1988), which outlines her Dominator and Partnership Culture thesis. The savvy memetic engineer is able to isolate, study, and subtly manipulate the underlying values systems, symbolic balance and primal atavisms that unconsciously influence the individual psyche and collective identity. A highly educated but susceptible intelligentsia, worldwide travel, and information vectors like the Internet, cable television, and tabloid media, means that hysterical epidemics and disinformation campaigns may become more common.

This warfare will be conducted using aesthetics, symbols, and doctrines as camouflage that will ultimately influence our cultural meme pool. These contemporary life conditions (Historic Times; Geographic Place; Existential Problems; and Societal Circumstances) are explored in books like Carl Sagan's *The Demon Haunted World: Science As a Candle in the Dark* (New York: Ballantine Books, 1996), John Brockman's *The Third Culture: Beyond the Scientific Revolution* (New York: Touchstone Books, 1996), and Michael Shermer's *Why People Believe Weird Things: Pseudo-science, Superstition,* and *Other Confusions of Our Time* (New York: W.H. Freeman & Co, 1996). Fictional descriptions of memetic engineering include Isaac Asimov's seminal Foundation Trilogy (New York: Bantam Books, 1991), George Gurdjieff's *Artificial Mythology Beelzebub's Tales to His Grandson* (New York: Penguin USA, 1999); Neil Stephenson's novels *Snow Crash* (New York: Bantam Spectra, 1993) and *The Diamond Age* (New York: Bantam Spectra, 1996); and Robert W. Chambers' unearthly *The King in Yellow* (Buccaneer Books, 1996) tome, which influenced seminal horror author H. P. Lovecraft.

Applied Memetic Engineering

Memetic engineering as a social science lends examples of itself in multiple areas and disciplines. It is currently being examined and researched by the US military as a means to counter insurgency and combat terrorism as explained below in *From the Clash to the Confluence of Civilizations* by Thomas P.M. Barnett an American military geostrategist, the Pentagon and Richard J. Pech's *Inhibiting Imitative Terrorism Through Memetic Engineering.*

Other examples of applied memetic engineering are present but not exclusive to the marketing and advertising industries. The question is whether these individuals can be truly considered memetic engineers. Marketing and advertising professionals create memes on an ongoing basis however, this alone doesn't necessarily qualify them as being memetic engineers.

Few if any actually fall into this category. This is possibly due to a lack of understanding of the various memes that have taken root in their target audience minds. According to the definition, industrial designers, music artist, athletes and other entertainers would more likely better serve this definition. This is because of their ability to create products, phrases and ideas that disseminate the population triggering a response within the brain causing a cultural phenomena. Game theory provides an empirical means of advancing the science of memetics. Memetic game theory, attempts to mathematically capture behavior in strategic situations; where an individual's success in making choices depends on the choices of others, based on past experiences, emotional behavior and learned behavior. It also offers a scientific approach to analysing social interactions.

According to G.Burchett and his research from G. Edward Griffin's book *The Creature from Jekyll Island: A Second Look at the Federal Reserve*, the concept of currency or money is considered a meme developed by early bankers. Here we have an example of a meme that was created to better serve the economic foundation of a society. This is a positive example of memetic engineering. However once the meme of legal tender was created it forced individuals to accept and use the currency under the guise of law. Once an individual is forced to accept currency by the government, essentially the government now controls the means of that individual's survival. In the case of the United States legal tender; the economic control of the citizens was placed in hands of the Federal Reserve which is a private entity.

Examples

An example of an engineered meme is Godwin's Law, a meme which propagates on mail-lists, and which its author professes to have initiated to reduce spam on those lists; one version is "When someone posts a metaphor about Nazis the thread is no longer useful." Richard Pech discusses the concept of memetic engineering within the context of countering mind contagions associated with terrorism. School shootings, for example, may be explained as an attempt to demonstrate the ultimate form of rebellion against a system in which the perpetrators feel ostracised or isolated. Acts of violence might appeal to their egos and the means for achieving this is replicated via the shooting meme. To re-engineer such a meme and its ability to infest susceptible minds, all information concerning such violence must be portrayed in an unappealing manner. For example, no one wants to be associated with acts of cowardice. By strongly suggesting that such violence is cowardly and the work of disturbed minds, it has less appeal for replication. In this manner the shooting meme has been re-engineered by removing its attraction, and therefore removing its ability to replicate.

Psycholinguistics

Psycholinguistics or psychology of language is the study of the psychological and neurobiological factors that enable humans to acquire, use, comprehend and produce language. Initial forays into psycholinguistics were largely philosophical ventures, due mainly to a lack of cohesive data on how the human brain functioned. Modern research makes use of biology, neuroscience, cognitive science, linguistics, and information theory to study how the brain processes language. There are a number of subdisciplines with non-invasive techniques for studying the neurological workings of the brain; for example, neurolinguistics has become a field in its own right. Psycholinguistics covers the cognitive processes that make it possible to generate a grammatical and meaningful sentence out of vocabulary and grammatical structures, as well as the processes that make it possible to understand utterances, words, text, etc. Developmental psycholinguistics studies children's ability to learn language.

Areas of Study

Psycholinguistics is an interdisciplinary field. Hence, it is studied by researchers from a variety of different backgrounds, such as psychology, cognitive science, linguistics, and speech and language pathology.

Psycholinguists study many different topics, but these topics can generally be divided into answering the following questions:

(1) How do children acquire language (language acquisition)?;

(2) How do people process and comprehend language (language comprehension)?;

(3) How do people produce language (language production)?; and

(4) How do adults acquire a new language (second language acquisition)? Subdivisions in psycholinguistics are also made based on the different components that make up human language.

Linguistics-related areas:

- Phonetics and phonology are concerned with the study of speech sounds. Within psycholinguistics, research focuses on how the brain processes and understands these sounds.
- Morphology is the study of word structures, especially the relationships between related words (such as dog and dogs) and the formation of words based on rules (such as plural formation).
- Syntax is the study of the patterns which dictate how words are combined to form sentences.
- Semantics deals with the meaning of words and sentences. Where syntax is concerned with the formal structure of sentences, semantics deals with the actual meaning of sentences.
- Pragmatics is concerned with the role of context in the interpretation of meaning.

A researcher interested in language comprehension may study word recognition during reading to examine the processes involved in the extraction of orthographic, morphological, phonological, and semantic information from patterns in printed text. A researcher interested in language production might study how words are prepared to be spoken starting from the conceptual or semantic level. Developmental psycholinguistics study infants' and children's ability to learn and process language.

Theories

Language Acquisition

There are essentially two schools of thought as to how children acquire or learn language, and there is still much debate as to which theory is the correct one. The first theory states that all language must be learned by the child. The second view states that the abstract system of language cannot be learned, but that humans possess an innate language faculty, or an access to what has been called universal grammar. The view that language must be learned was especially popular before 1960 and is well represented by the mentalistic theories of Jean Piaget and the empiricist Rudolf Carnap. Likewise, the school of psychology known as behaviourism by B.F. Skinner puts forth the point of view that language is a behaviour shaped by conditioned response, hence it is learned.

The innatist perspective began with Noam Chomsky's highly critical review of Skinner's book in 1959. This review helped to start what has been termed "the cognitive revolution" in psychology. Chomsky posited humans possess a special, innate ability for language and that complex syntactic features, such as recursion, are "hard-wired" in the brain. These abilities are thought to be beyond the grasp of the most intelligent and social non-humans. According to Chomsky, children acquiring a language have a vast search space to explore among all possible human grammars, yet at the time there was no evidence that children receive sufficient input to learn all the rules of their language. Hence, there must be some other innate mechanism that endows a language ability to humans. Such a language faculty is, according to the innatist theory, what defines human language and makes it different from even the most sophisticated forms of animal communication.

The field of linguistics and psycholinguistics since then has been defined by reactions to Chomsky, pro and con. The pro view still holds that the human ability to use language (specifically the ability to use recursion) is qualitatively different from any sort of animal ability. This ability may have resulted from a favorable mutation or from an adaptation of skills evolved for other purposes. The view that language can be learned has had a recent resurgence inspired by emergentism. This view challenges the "innate" view as scientifically unfalsifiable; that is to say, it can't be tested. With the amount of computer power increasing since the 1980s, researchers have been able to simulate language acquisition using neural network models. These models provide evidence that there may, in fact, be sufficient information contained in the input to learn language, even syntax. If this is true, then an innate mechanism is no longer necessary to explain language acquisition.

Language Comprehension

One question in the realm of language comprehension is how people understand sentences as they read (also known as sentence processing). Experimental research has spawned a number of theories about the architecture and mechanisms of sentence comprehension. Typically these theories are concerned with what types of information contained in the sentence the reader can use to build meaning, and at what point in reading does that information become available to the reader. Issues such as "modular" versus "interactive" processing have been theoretical divides in the field.

A modular view of sentence processing assumes that the stages involved in reading a sentence function independently in separate modules. These modulates have limited interaction with one another. For example, one influential theory of sentence processing, the garden-path theory, states that syntactic analysis takes place first. Under this theory as the reader is reading a sentence, he or she creates the simplest structure possible in order to minimize effort and cognitive load. This is done without any input from semantic analysis or context-dependent information. Hence, in the sentence "The evidence examined by the lawyer turned out to be unreliable," by the time the reader gets to the word "examined" he or she has committed to a reading of the sentence in which the evidence is examining something because it is the simplest parse. This commitment is made despite the fact that it results in an implausible situation; we know that experience that evidence can rarely if ever examine something. Under this "syntax first" theory, semantic information is processed at a later stage. It is only later that the reader will recognize that her or she needs to revise the initial parse into one in which "the evidence" is being examined. In this example, readers typically recognize their misparse by the time they reach "by the lawyer" and must go back and re-parse the sentence. This reanalysis is costly and contributes to slower reading times.

In contrast to a modular account, an interactive theory of sentence processing, such as a constraint-based lexical approach assumes that all available information contained within a sentence can be processed at any time. Under an interactive account, for example, the semantics of a sentence (such as plausibility) can come into play early on in order to help determine the structure of a sentence. Hence, in the sentence above, the reader would be able to make use of plausibility information in order to assume that "the evidence" is being examined instead of doing the examining. There are data to support both modular and interactive accounts; which account is the correct one is still up for debate.

Methodologies

Behavioural Tasks

Many of the experiments conducted in psycholinguistics, especially earlier on, are behavioural in nature. In these types of studies, subjects are presented with linguistic stimuli and asked to perform an action. For example, they may be asked to make a judgement about a word (lexical decision), reproduce the stimulus, or name a visually presented word aloud). Reaction times to respond to the stimuli(usually on the order of milliseconds) and proportion of correct responses are the most often employed measures of performance in behavioural tasks. Such experiments often take advantage of priming effects, whereby a "priming" word or phrase appearing in the experiment can speed up the lexical decision for a related "target" word later.

As an example of how behavioural methods can be used in psycholinguistics research, Fischler (1977) investigated word encoding using the lexical decision task. She asked participants to make decisions about whether two strings of letters were English words. Sometimes the strings would be actual English words requiring a "yes" response, and other times they would be nonwords requiring a "no" response. A subset of the licit words were related semantically (e.g., cat-dog) while others were unrelated (e.g., bread-stem). Fischler found that related word pairs were responded to faster when compared to unrelated word pairs. This facilitation suggests that semantic relatedness can facilitate word encoding.

Eye-movements

Recently, eye tracking has been used to study online language processing. Beginning with Rayner (1978) the importance and informativity of eye-movements during reading was established. Later, Tanenhaus et al. (1995) used the visual-world paradigm to study the cognitive processes related to spoken language. Assuming that eye movements are closely linked to the current focus of attention, language processing can be studied by monitoring eye movements while a subject is presented auditorily with linguistic input.

Neuroimaging

Until the recent advent of non-invasive medical techniques, brain surgery was the preferred way for language researchers to discover how language works in the brain. For example, severing the corpus callosum (the bundle of nerves that connects the two hemispheres of the brain) was at one time a treatment for some forms of epilepsy. Researchers could then study the ways in which the comprehension and production of language were affected by such drastic surgery. Where an illness made brain surgery necessary, language researchers had an opportunity to pursue their research.

Newer, non-invasive techniques now include brain imaging by positron emission tomography (PET); functional magnetic resonance imaging (fMRI); event-related potentials (ERPs) in electroencephalography (EEG) and magnetoencephalography (MEG); and transcranial magnetic stimulation (TMS). Brain imaging techniques vary in their spatial and temporal resolutions (fMRI has a resolution of a few thousand neurons per pixel, and ERP has millisecond accuracy). Each type of methodology presents a set of advantages and disadvantages for studying a particular problem in psycholinguistics.

Computational Modeling

Computational modeling—e.g. the DRC model of reading and word recognition proposed by Coltheart and colleagues—is another methodology. It refers to the practice of setting up cognitive models in the form of executable computer programs. Such programs are useful because they require theorists to be explicit in their hypotheses and because they can be used to generate accurate predictions for theoretical models that are so complex that they render discursive analysis unreliable. Another example of computational modeling is McClelland and Elman's TRACE model of speech perception.

Issues and Areas of Research

Psycholinguistics is concerned with the nature of the computations and processes that the brain undergoes to comprehend and produce language. For example, the cohort model seeks to describe how words are retrieved from the mental lexicon when an individual hears or sees linguistic input.

Recent research using new non-invasive imaging techniques seeks to shed light on just where certain language processes occur in the brain.

There are a number of unanswered questions in psycholinguistics, such as whether the human ability to use syntax is based on innate mental structures or emerges from interaction with other humans, and whether some animals can be taught the syntax of human language.

Two other major subfields of psycholinguistics investigate first language acquisition, the process by which infants acquire language, and second language acquisition. In addition, it is much more difficult for adults to acquire second languages than it is for infants to learn their first language (bilingual infants are able to learn both of their native languages easily). Thus, sensitive periods may exist during which language can be learned readily. A great deal of research in psycholinguistics focuses on how this ability develops and diminishes over time. It also seems to be the case that the more languages one knows, the easier it is to learn more.

The field of aphasiology deals with language deficits that arise because of brain damage. Studies in aphasiology can both offer advances in therapy for individuals suffering from aphasia, and further insight into how the brain processes language.

Spiral Dynamics

Spiral Dynamics is a theory of human development introduced in the 1996 book *Spiral Dynamics* by Don Beck and Chris Cowan. The book was based on the theory of psychology. Professor Clare W. Graves "Spiral Dynamics" is a registered trademark of the National Values Center, Inc.

Overview

Spiral Dynamics argues that human nature is not fixed: humans are able, when forced by life conditions, to adapt to their environment by constructing new, more complex, conceptual models of the world that allow them to handle the new problems. Each new model transcends and includes all previous models. According to Beck and Cowan, these conceptual models are organized around so-called Memes: systems of core values or collective intelligences, applicable to both individuals and entire cultures.

In spiral dynamics, the term Meme refers to a core value system, acting as an organizing principle, which expresses itself through memes (self-propagating ideas, habits, or cultural practices). The prepended and superscripted letter v indicates these are not basic memes but value systems which include them. The colours act as reminders for the Life Conditions and Mind Capacities of each system and alternate between cool and warm colors as a part of the model. Within the model, individuals and cultures do not fall clearly in any single category (colour). Each person/culture embodies a mixture of the value patterns, with varying degrees of intensity in each. Spiral Dynamics claims not to be a linear or hierarchical model, although this assertion has been contested.

According to Spiral Dynamics, there are infinite stages of progress and regression over time dependent upon the life circumstances of the person/culture, which are constantly in flux. Attaining higher stages of development is not synonymous with attaining a 'better' or 'more correct' values system. All stages co-exist in both healthy and unhealthy states, whereby any stage of development can lead to undesirable outcomes with respect to the health of the human and social environment.

Memes

First tier Meme levels are focused on different themes for existence, and include almost all of the worldviews, cultures, and mental attitudes up to today. New systems build on adaptations of previous levels and seek to solve problems created by living in those earlier ways. (Attaching concrete examples to these levels of psychological existence is difficult and often misleading because (a) there can be multiple reasons for the same behaviour and (b) centralization in a single level regarding all aspects of living is rare. Some advocates of Spiral Dynamics argue that these are ways of thinking about things, not types of people.

However, Spiral Dynamics does explicitly define a category of people labeled "Spiral Wizards" who have attained equilibrium around "higher level" Memes.

Beige

- Summary: Archaic-instinctive—survivalistic/automatic/reflexological
- Time of origin: c. 100,000 BC
- Description: "Express self to meet imperative physiological needs through instincts of Homo sapiens."

Purple

- Summary: Animistic-tribalistic magical-animistic Tribal order
- Time of origin: 50,000 BC
- Description: "Sacrifice to the ways of the elders and customs as one subsumed in group." This is the level of traditional cultures.

Red

- Summary: Egocentric-exploitive power gods/dominionist
- Time of origin: 7000 BC
- Description: "Express self (impulsively) for what self desires without guilt and to avoid shame." Expressed by the mentality of street gangs, Vikings, etc.

Blue

- Summary: Absolutistic-obedience mythic order—purposeful/authoritarian
- Time of origin: c. 3000 BC
- Description: "Sacrifice self for reward to come through obedience to rightful authority in purposeful Way." Embodied by fundamentalist religions.
- Note: Amber is Ken Wilber's current name for this stage.

Orange

- Summary: Multiplistic-achievist scientific/strategic
- Time of origin: c. 1000 AD on (as early as 600 AD according to Graves and Calhoun)
- Description: "Express self (calculatedly) to reach goals and objectives without rousing the ire of important others." Expressed in the Scientific Revolution and the Industrial Revolution.

Green

- Summary: Relativistic-personalistic—communitarian/egalitarian
- Time of origin: From 1850 on (surged in early 20th century)
- Description: "Sacrifice self-interest now in order to gain acceptance and group harmony." Expressed in 1960s pluralism and systems theory.

Spiral Dynamics theory sees second tier memes as emerging levels that gradually move away from a focus on subsistence-level concerns of the First Tier, and towards a being-level existence. The existence of two different tiers of psychological development was introduced by the founder Graves. Cowan claims it is possible Graves introduced the tier system as a marketing instrument. Up to today there is no research evidence the two tiers exist. Cowan no longer supports the existence of two tiers but claims the only thing now known about the Yellow and Turquoise systems is that they are more complex versions of Orange and Green. The open-ended theory suggests that the levels Coral and beyond are not yet substantially formed and will solidify as a greater portion of society develops towards those memes.

Yellow

- Summary: Systemic-integrative
- Time of origin: 1950s
- Description: "Express self for what self desires, but to avoid harm to others so that all life, not just own life, will benefit."

Turquoise

- Summary: Holistic
- Time of origin: 1970s
- Description: An integrative system which combines an organism's necessary self-interest with the interests of the communities in which it participates. The theory is still forming.

Development

The colour system was added in the 1970s as a graphic element to decorate training materials used by Cowan and Beck. The term Meme was introduced by Beck and Cowan in Spiral Dynamics where the colour language replaced original Graves terminology. Graves had used letter pairs to refer to each level and had not considered any connection with memetics. Beck and Cowan emphasized 'change states' which are part of the Graves theory. They identify landmarks on the transformational path between the levels. Graves' original theory uses a double helix model to show the interrelatedness of an individual's perception of life conditions with their inner neuronal systems, producing a level of psychological existence. This double helix of two interacting forces is referred to as a spiral in 'Spiral Dynamics.

Following the release of their book, Beck and Cowan taught this theory in two courses, SDI and SDII. They ceased their formal working relationship in 1999. Since this break both have continued to offer training courses of various types as well as carrying out consultancy work. Beck spent several years applying the theory of Spiral Dynamics while working with individuals and groups in South Africa involved in the transition from Apartheid for which he received a legislative honour from the state of Texas. Beck has continued to facilitate sessions using the model in both business and politics. Beck and Ken Wilber (with his integral theory) became interested in each other's work, resulting in Beck developing a branch of spiral dynamics that he calls Spiral Dynamics Integral. This version of the theory uses integral concepts such as the four quadrants. Beck is also associated with the spiritual teacher, Andrew Cohen. This approach continues to be applied and developed in organisational and societal contexts.

Cowan does not subscribe to these developments and promotes a version of the theory which he describes as remaining more faithful to the original research of Clare Graves and extending from it. He continues to use the term 'Spiral Dynamics' to describe his work since he co-created it. With his partner, Natasha Todorovic, he has undertaken work in integrating Spiral Dynamics with NLP and other models, and in developing corporate strategy and practical applications. He has also suggested and additional second tier meme Coral which deals with the development of neurological capacities Each of the external websites listed below promotes the organization of one of the two co-authors.

Pathologies

According to Don Beck and Ken Wilber, each Meme has both healthy and unhealthy versions. The pathologies are sometimes referred to as being "mean" as in "Mean Green Meme" (MGM) or "Mean Orange Meme" (MOM) . As examples, the MOM includes the extremes of capitalism like exploitation, environmental devastation and a general lack of ethics and sensitivity, while the MGM includes performative contradictions like anti-hierarchy, anti-competition, etc. Cowan disputes that any credible evidence exists for the existence of the "Mean Green Meme" and that it is a misrepresentation of the theory. While he recognizes the problems mentioned above, Cowan argues that they exist in other value systems as well, such that attaching them only to the Green Meme is too simplistic. He considers the term "mean" to be inappropriate and a theoretical distortion when questions of adaptation or maladaptation, congruence or ineffectiveness are more to the point. Psychopathology potentially exists at all levels and is a different dimension.

Todorovic offers data and analysis leading to the conclusion that the "Mean Green Meme" conjecture is contradicted by the data. Todorovic further concludes that MGM is a "failure of analysis" based on a variety of misconceptions about spiral dynamics and constitutes an "alarming misdiagnosis" and brands MGM a harmful "form of spiral fundamentalism."

Criticisms of Spiral Dynamics

Critics point out that the model's implications are political as well as developmental and that while the terminology of the theory is self-consciously inclusive, the practical implications of the model can be seen as socially elitist and authoritarian. In his work on the subject, Beck emphasizes that one of the characteristics of "tier two" individuals, also called "Spiral Wizards", is their ability to make superior decisions for all parties concerned and to manufacture consent for their approaches at lower levels using resonant terms and ideas. In addition to outlining an underlying developmental theory, Spiral Dynamics gives explicit suggestions to these "Wizards" for both consensual and non-consensual management of "lower-tier" individuals. One critic of Spiral Dynamics, Michel Bauwens, has argued that some conceptions of what it means to be "second tier" have come to resemble Nietzsche's idea of the Übermensch. Co author Cowan has publicly dissociated himself from the ideas that are currently being promoted by his ex-partner Beck in conjunction with Ken Wilber.

Predictive Medicine

Predictive medicine is a rapidly emerging field of medicine that entails predicting disease and instituting preventive measures in order to either prevent the disease altogether or significantly decrease its impact upon the patient (such as by preventing mortality or limiting morbidity). While different prediction methodologies exist, such as genomics, proteomics, and cytomics, the most fundamental way to predict future disease is based on genetics.

Although proteomics and cytomics allow for the early detection of disease, much of the time those detect biological markers that exist because a disease process has already started. However, comprehensive genetic testing (such as through the use of DNA arrays or full genome sequencing) allows for the estimation of disease risk years to decades before any disease even exists, or even whether a healthy fetus is at higher risk for developing a disease in adolescence or adulthood. Individuals who are more susceptible to disease in the future can be offered lifestyle advice or medication with the aim of preventing the predicted illness. Current genetic testing guidelines supported by the health care professionals discourage purely predictive genetic testing of minors until they are competent to understand the relevancy of genetic screening so as to allow them to participate in the decision about whether or not it is appropriate for them. Genetic screening of newborns and children in the field of predictive medicine is deemed appropriate if there is a compelling clinical reason to do so, such as the availability of prevention or treatment as a child that would prevent future disease.

5

The Cell Basic Applications

The cell is one of the most basic units of life. There are millions of different types of cells. There are cells that are organisms onto themselves, such as microscopic amoeba and bacteria cells. And there are cells that only function when part of a larger organism, such as the cells that make up your body. The cell is the smallest unit of life in our bodies. In the body, there are brain cells, skin cells, liver cells, stomach cells, and the list goes on.

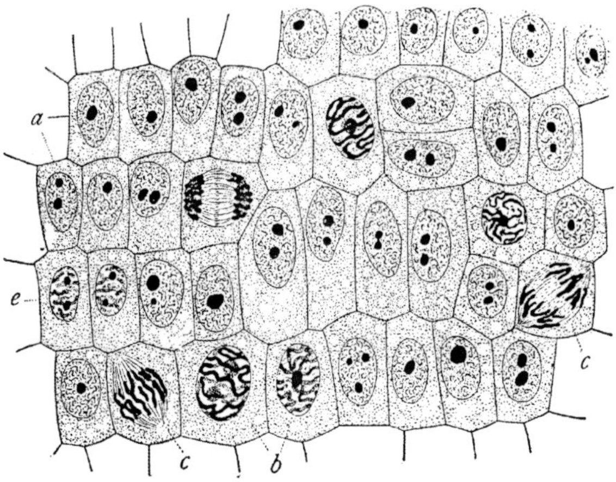

***Figure:** Allium cells in different phases of the cell cycle*

All of these cells have unique functions and features. And all have some recognizable similarities. All cells have a 'skin', called the plasma membrane, protecting it from the outside environment. The cell membrane regulates the movement of water, nutrients and wastes into and out of the cell. Inside of the cell membrane are the working parts of the cell. At the centre of the cell is the cell nucleus.

The cell nucleus contains the cell's DNA, the genetic code that coordinates protein synthesis. In addition to the nucleus, there are many organelles inside of the cell-small structures that help carry out the day-to-day operations of the cell. One important cellular organelle is the ribosome. Ribosomes participate in protein synthesis.

The transcription phase of protein synthesis takes places in the cell nucleus. After this step is complete, the mRNA leaves the nucleus and travels to the cell's ribosomes, where translation occurs. Another important cellular organelle is the mitochondrion. Mitochondria (many mitochondrion) are often referred to as the power plants of the cell because many of the reactions that produce energy take place in mitochondria. Also important in the life of a cell are the lysosomes. Lysosomes are organelles that contain enzymes that aid in the digestion of nutrient molecules and other materials. Below is a labelled diagram of a cell to help you identify some of these structures.

There are many different types of cells. One major difference in cells occurs between plant cells and animal cells. While both plant and animal cells contain the structures, plant cells have some additional specialized structures. Many animals have skeletons to give their body structure and support. Plants do not have a skeleton for support and yet plants don't just flop over in a big spongy mess. This is because of a unique cellular structure called the cell wall.

The cell wall is a rigid structure outside of the cell membrane composed mainly of the polysaccharide cellulose. As pictured at left, the cell wall gives the plant cell a defined shape which helps support individual parts of plants. In addition to the cell wall, plant cells contain an organelle called the chloroplast. The chloroplast allow plants to harvest energy from sunlight. Specialized pigments in the chloroplast (including the common green pigment chlorophyll) absorb sunlight and use this energy to complete the chemical reaction:

$$6\ CO_2 + 6\ H_2O + \text{energy (from sunlight)} \rightarrow C_6H_{12}O_6 + 6\ O_2$$

In this way, plant cells manufacture glucose and other carbohydrates that they can store for later use.

Organisms contain many different types of cells that perform many different functions. In the next lesson, we will examine how individual cells come together to form larger structures in the human body.

Each cell has a part called a cell membrane; a thin covering around the cell. The cell membrane separates it, gives the cell its shape, and controls what goes into and out of it. Located inside a cell is the cytoplasm. Cytoplasm is a jellylike liquid. Inside the cytoplasm is the cell's control centre called the nucleus. The nucleus contains all of the instructions for running the cell. The nucleus is surrounded by the nuclear membrane. The nuclear membrane controls what goes into and comes out of the nucleus.

A cell is able to reproduce by means of splitting. One cell splits into two cells. However before the cell splits, the nucleus doubles and divides in half so each new cell has a copy of the cell's instructions.

Look carefully at the diagram of the cell. Notice inside the cytoplasm are vacuoles and mitochondria. Vacuoles are tiny oval structures that store food, water, or wastes. Mitochondria are shaped like kidney beans. Mitochondria are important parts of a cell because they help change food into energy. The cell uses the energy to do its work.

Cells do not all look the same. Their shape depends upon the jobs they perform. Red blood cells are tiny and round. They need to be so they can squeeze through blood vessels and bring oxygen to the other cells in the body. Muscle cells are thin and long. This is so they can expand and contract to help the body move.

All cells, whether plant or animal have these same parts: cell membrane, cytoplasm, nucleus, nuclear membrane, vacuoles, and mitochondria. Plant cells however have two additional parts that animal cells do not. Plant cells have cell walls. This sturdy layer around the cell membrane supports and protects the cell. Plant cells also contain chloroplasts. Chloroplasts contain the green substance called chlorophyll. Chlorophyll traps energy from sunlight and enables a plant to make its own food.

The Cell Cycle

All organisms are composed of cells. Organisms may be either single celled, as found in the Kingdoms of Archaebacteria and Eubacteria and Protista; or multicellular, as found in the Kingdoms Fungi, Plantae and Animalia. While all of the organisms found on Earth are quite diverse, they all have several characteristics in common:

- the ability to metabolize, or maintain the biochemical processes necessary for life
- the ability to grow and evolve
- the ability to reproduce.

All of the organisms in the Kingdom Archaebacteria and Eubacteria are prokaryotic bacteria. These single celled organisms have no nucleus and no complex membrane bound organelles. They reproduce by an asexual process called binary fission. Binary fission means literally "splitting into two," and involves no genetic recombination. Thus, this process results in the formation of identical cells, or clones. Because this is true for all bacteria you may wonder how different strains of bacteria arise. For example, how does non-pathogenic *Escherichia coli* (*E. coli*), identical to that found in your gut somehow produce killer *E. coli* strain O157:H7, which caused many deaths in the western United States in 1993 and continues to pose a threat? The many variations within a given species of bacteria arise from mutation of its DNA by either mutagenic agents such as UV light or by viral infection of the bacterial cell.

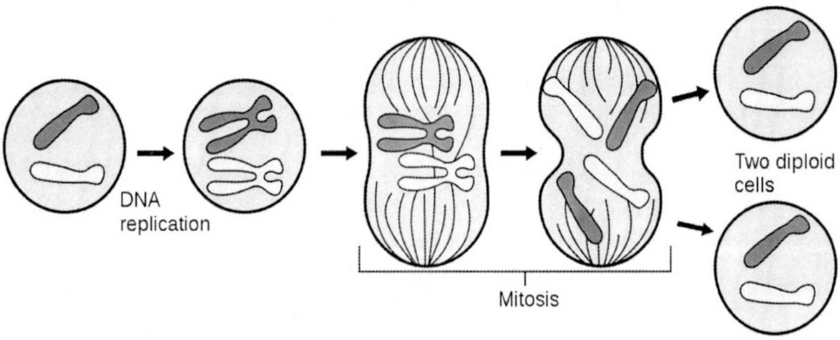

Figure: *Each turn of the cell cycle divides the chromosomes in a cell nucleus.*

While mutagenic agents and viruses can also alter eukaryotic cells, the effect is generally not seen in the offspring. The primary force contributing to genetic diversity in the multicellular eukaryotes is sexual reproduction. Sexual reproduction involves the recombination of genetic material between two different individuals of the same species to produce a genetically unique offspring. For example, you obtained one half of your genetic information (DNA) from your mother and one half from your father-thus you are a combination of your two parents and quite genetically distinct from all other humans (unless you have an identical twin). You came into being when two haploid gametes (sperm and egg) fused and became a single diploid cell, called a zygote.

Since you started as a single cell, how did you become the complex multicellular organism you are today? The zygote's first job is to divide and produce many identical cells, through the processes of mitosis and cytokinesis. Mitosis involves the division of the nucleus, and cytokinesis involves the division of the cytoplasm. This results in the formation of two identical cells from one cell.

How does mitosis and cytokinesis occur, and what does the cell do between these division events? Both of these processes occur only after cell growth, metabolism, and DNA duplication within the nucleus of the cell. Cells within a multicellular organism have a distinct life span, which is called the cell cycle. Consider the cell cycle as the life span of a single cell.

The cell is "born" after cell division, it grows and metabolizes in a cell cycle phase called G_1 phase, and then it prepares to reproduce. First, it must duplicate all of its genetic information, or DNA, in a cell cycle phase called S phase. Then it must spend some time metabolizing all the necessary materials that it needs to split into two cells during the phase called G_2 phase. Finally the nucleus of the cell, which now has double the amount of its DNA, goes through nuclear division. This nuclear division is called mitosis. When mitosis is almost complete, the cell cytoplasm must split. The formation of new cell wall or membrane then results in two identical cells.

Each type of cell, whether epithelial, cardiac, nerve, or some other type, has its own distinct cell cycle, so the time spent in each phase may vary between different cell types. However, all eukaryotic cells in multicellular organisms have the same generalized cell cycle containing G_1, S, G_2, Mitosis, and Cytokinesis phases.

Homologous Chromosomes, Genes, and DNA

Almost everyone in the media talks about chromosomes, DNA, and genes as if they were all the same entity. While they are related, one must realize the relationship between these terms to gain a real understanding of how the most basic of life processes occur. Chromosomes are highly condensed (tightly wound), huge biomolecules, composed of DNA, protein, and some RNA. Some organisms contain only one unpaired chromosome that is not contained in a nuclear envelope. All of these organisms are called prokaryotes.

Eukaryotes, by contrast, especially in multicellular organisms, have several chromosomes per cell. These chromosomes exist as paired, or homologous, chromosomes. The homologous chromosomes condense and pair during the mitosis stage of the life cycle of all eukaryotes. The reason these chromosomes are paired is that half of an organism's chromosomes came from the paternal gamete and the other half came from the maternal gamete.

Let us use human beings again as an example. Humans have 46 individual chromosomes in every cell of their body during mitosis, except for the gametes (sex cells like eggs or sperm) and mature red blood cells. However, there are only 24 different human chromosomes. How can this be?

Humans acquire a set of 22 chromosomes and one X chromosome from their mother for a total of 23, and a set of 22 chromosomes and one X or Y chromosome from their father to make up the 46 chromosomes. Does the mother's chromosome number five differ from the father's chromosome number five? Yes! The two chromosomes do differ in their specific information. While both homologous chromosomes contain the same type of information located at the same exact place, they each have a different version of the information.

The chromosome from the one parent may contain information for blue eye colour, and the chromosome from the other parent may contain information for brown eye colour. While the specific DNA sequence information may vary from individual to individual, chromosome number five in humans encodes for the same type of product.

The pairs of homologous chromosomes contain redundant information so that you have two copies of each chromosome, one from Mom and one from Dad. In this way, you may exhibit traits from both parents.

As you probably know, human males have two different sex-determining chromosomes: one X and one Y. These are not homologous, but contain very different information. Females humans have two X chromosomes. So besides the regular 22 chromosomes, there are also two different sex chromosomes, for a total of 24 different chromosomes in humans.

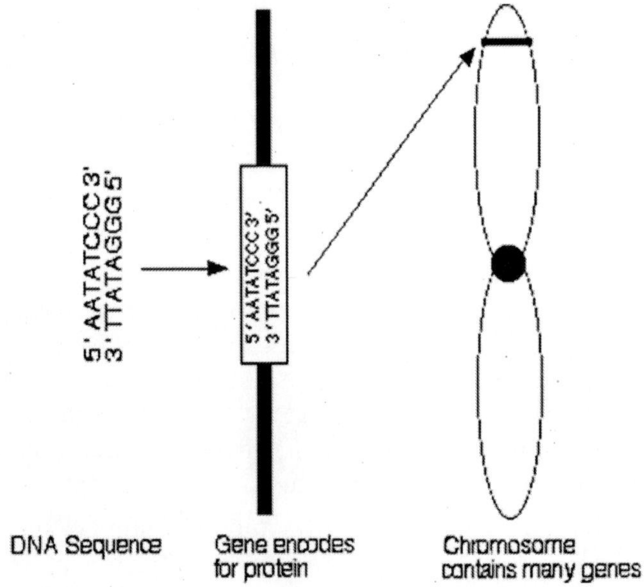

***Figure:** The relationship between DNA, genes, and chromosomes.*

How exactly do genes relate to chromosomes? First you must realize that DNA, the essential molecule of the chromosome, contains all the information of the cell for the manufacture of all proteins necessary to maintain and perpetuate life. A gene is a segment of DNA that encodes for one polypeptide, or molecule of protein. The colour of your eyes results from the production of specific polypeptides, which come together to form a protein that exhibits physical properties that makes them appear to have a certain colour. These polypeptides are coded by genes or a set of genes. The genes are DNA sequences contained in a chromosome. There are many genes on a single chromosome. Note that much of the DNA in a chromosome does not encode for a polypeptide and is not gene DNA. The function of non-genic DNA is not well understood, although some parts of this DNA are known to regulate the expression of genes.

Mitosis: Division of the Cell Nucleus

Now to discuss how the process of mitosis occurs in a cell. It is very important to remember that the cell must have duplicated all of its DNA in S phase of the cell cycle, before mitosis can occur. If an organism has three pairs of homologous chromosomes, how many individual chromosomes does it have? How many chromosomes does this same cell have after S phase of the cell cycle?

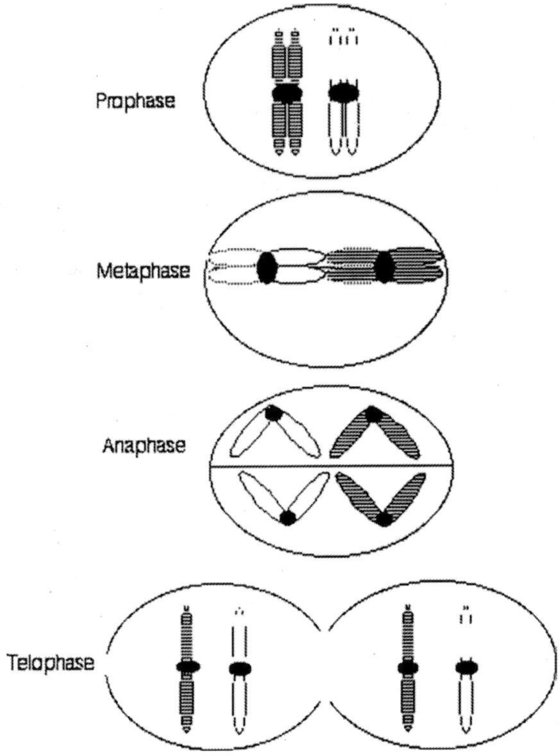

Figure: *Representation of a cell going through mitosis.*

Let us discuss the terms diploid and haploid. Any organism or cell that has paired, homologous chromosomes is diploid. If a cell or organism has only one chromosome of each type it is considered haploid. Ploidy does not deal with the numbers of chromosomes, but instead deals with having paired or unpaired chromosomes. Most cells in your body are diploid and contain two of each type of chromosome. Gametes, or sex cells, only contain one set of chromosomes, so they contain half the diploid number. Gametes are considered haploid cells, having only one of each of the homologous chromosomes. Now to answer the above questions: an organism having three pairs of homologous chromosomes has six individual chromosomes. After S phase when the DNA replicates, the cell contains 12 individual chromosomes. Now the cell can form two nuclei, each having one pair of each type of chromosome, and the cell can split to form two identical cells.

Mitosis occurs in several stages. Remember that these stages are part of a very rapid, continuous process and are not discrete events as might seem from their descriptions. It is important to learn the names of each stage and to be able to describe the cellular processes that occur.

Outline of stages of Mitosis and Cytokinesis:

I. Mitosis:

   a. Prophase

   b. Metaphase

   c. Anaphase

   d. Telophase

II. Cytokinesis.

The most important information to learn about Mitosis is the position and number of chromosomes in each of the stages of Mitosis. You must be able to answer a question similar to the following example: diagram a cell that has 2 pairs of homologous chromosomes in each of the stages of mitosis.

Prophase: The first stage of Mitosis when chromosomes condense and the nuclear membrane disappears. The chromosomes that have duplicated in S phase attach to their duplicate in pairs at a central point called a centromere. At this stage the attached chromosomes are called sister chromatids. Sister chromatids are identical and are very different from homologous chromosomes. All these sisters are identical twins. Each homologous chromosome has been duplicated, and now the two duplicates are attached to one another to form sister chromatids.

Metaphase: The stage at which the chromosomes line up at the equator of the cell and spindle fibres attach to the centromere of each chromosome. The spindle fibres, made of protein subunits, shorten and start to pull the chromosomes to opposite sides of the cell.

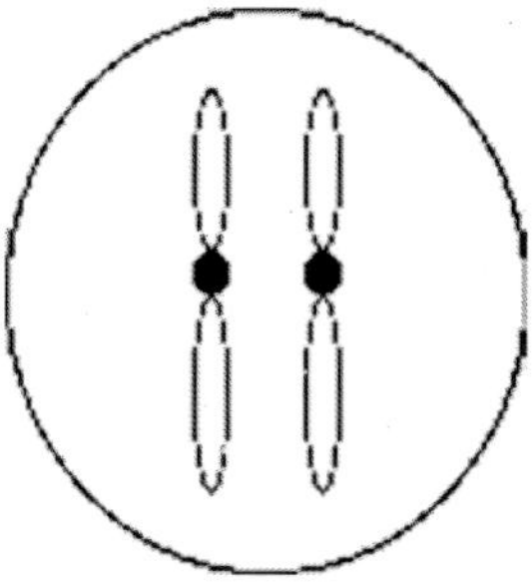

Figure: *Diploid cell containing one pair of homologous chromosomes before S phase of the cell cycle. This is what a normal diploid cell would look like if chromosomes were condensed before S phase occurred (they aren't actually condensed until prophase) and what both cells should look like immediately following telophase, before the chromosomes decondense.*

Anaphase: The stage at which sister chromatids (identical duplicates) are separated and pulled to opposite poles of the cell.

Telophase: The stage at which the cell membrane or cell wall forms in the centre of the cell.

Cytokinesis: The stage where the cytoplasm and organelles are divided between the two cells and the chromosomes decondense, forming two diploid, identical cells.

Meiosis: the Formation of Gametes

We have previously discussed the concept of sexual reproduction. Generally, in most organisms the chromosome content of the cell must be diploid, or 2N, where N represents the haploid number of chromosomes.

In other words, cells usually do not tolerate more than two homologous chromosomes in the cell at one time. For example, if a human zygote were formed by the fusion of two diploid cells, rather than two haploid gamete cells, the zygote would have 92 chromosomes. That translates to four sets of the 22 chromosomes and four sex chromosomes. You can see that this would soon become a huge problem for the cell!

To avoid these problems, most organisms donate only one half of a set of chromosomes during sexual recombination. The resulting zygote will therefore have exactly the same number of chromosomes as the parents. The process by which the cell accomplishes this feat is called Meiosis. Meiosis is a type of cell division that occurs only in diploid cells destined to become gametes. No other type of cell undergoes Meiosis.

Before Meiosis begins, the diploid cell goes through the entire cell cycle including S phase, just exactly as a cell does before Mitosis occurs. Since the cell now has double the number of chromosomes of the original diploid cell, it must go through two rounds of division to become haploid.

Example: A cell that contains two pairs of homologous chromosomes, or four chromosomes in all, goes through S phase before Meiosis and now has four pairs of homologous chromosomes, or eight chromosomes in all. The first meiotic division results in two cells each having two pairs of homologous chromosomes. The second meiotic division, undergone by both cells, results in four cells each having two chromosomes, one of each homologous pair.

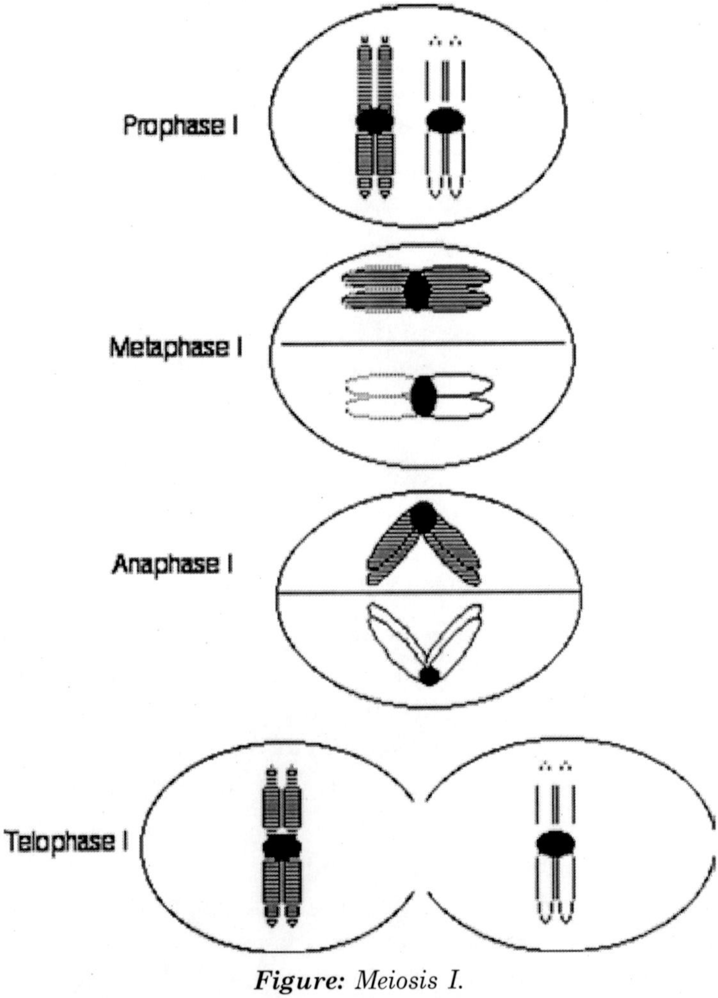

Figure: *Meiosis I.*

The process of Meiosis has many similarities to Mitosis. Again, you should write a brief description for each of the following phases of Mitosis.

I. Meiosis I

a. Prophase I

b. Metaphase I

c. Anaphase I

d. Telophase I

II. Cytokinesis

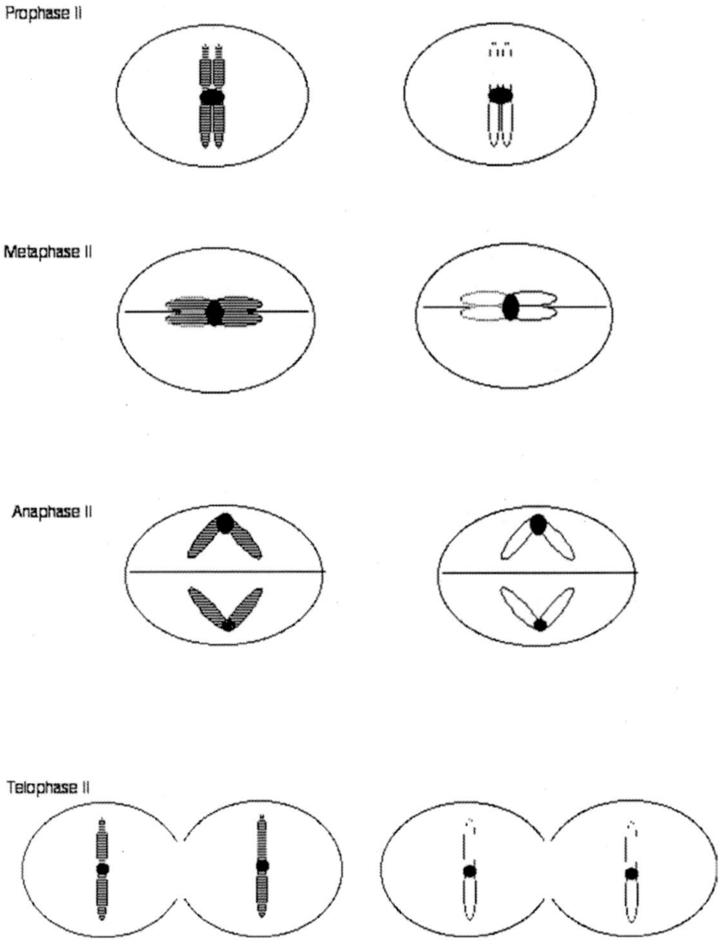

Figure: *Meiosis II.*

III. Meiosis II

a. Prophase II

b. Metaphase II

c. Anaphase II

d. Telophase II

IV. Cytokinesis

The most important information for you to know about Meiosis is the position and number of chromosomes in each of the stages. You must be able to answer a question similar to the following example: Please diagram a cell which has two pairs of homologous chromosomes in each of the stages of Meiosis.

Prophase I: The first stage of Meiosis when chromosomes condense and the nuclear membrane disappears. The chromosomes that have duplicated in S phase attach to their duplicate in pairs at a central point called a centromere. At this stage the attached chromosomes are called sister chromatids. Sister chromatids are identical and are very different from homologous chromosomes. All these sisters are identical twins. Each homologous chromosome has been duplicated, and now the two duplicates are attached to one another to form sister chromatids.

Metaphase I: The stage at which the chromosomes line up at the equator of the cell and spindle fibres attach to the centromere of each chromosome. The spindle fibres, made of protein subunits, shorten and start to pull the chromosomes to opposite sides of the cell.

Anaphase I: The stage at which sister chromatids (identical duplicates) stay together and are pulled to opposite poles of the cell.

Telophase I: The stage at which the cell membrane or cell wall forms in the centre of the cell.

Cytokinesis: The stage where the cytoplasm and organelles are divided between the two cells and the chromosomes decondense, forming two diploid, identical cells.

And now for Meiosis II...

Prophase II: The stage of Meiosis II when chromosomes condense and the nuclear membrane disappears. The chromosomes have not been duplicated, and the cell is still only diploid.

Metaphase II: The stage at which the chromosomes line up at the equator of the cell and spindle fibres attach to the centromere of each attached sister chromatid.

Anaphase II: The stage at which sister chromatids separate and are pulled to opposite poles of the cell.

Telophase II: The stage at which the cell membrane or cell wall forms in the centre of the cell.

Cytokinesis: The stage where the cytoplasm and organelles are divided between the two cells and the chromosomes decondense, forming two haploid gamete cells. The gamete has one half of the information of a diploid cell. Each of the four gametes produced has one of each homologous chromosome, in contrast to a diploid cell that has homologous pairs of chromosomes.

Enzymes: The Catalysts of Life

Each cell in the body is a beehive of activity called metabolism. The biochemical activity is the process of chemical and physical change which goes on continually in the living organism. These characteristics of life include the build-up of new tissue, replacement of old tissue, conversion of food to energy, disposal of waste materials, and reproduction.

This building up and tearing down takes place in the face of an apparent paradox. The greatest majority of these biochemical reactions do not take place spontaneously. The phenomenon of catalysis makes possible biochemical reactions necessary for all life processes. Catalysis is defined as the acceleration of a chemical reaction by some substance which itself undergoes no permanent chemical change. The catalysts of biochemical reactions are enzymes and are responsible for bringing about almost all of the chemical reactions in living organisms. Without enzymes, these reactions take place at a rate far too slow for the pace of metabolism.

The oxidation of a fatty acid to carbon dioxide and water is not a gentle process in a test tube – extremes of pH, high temperatures and corrosive chemicals are required. Yet in the body, such a reaction takes place smoothly and rapidly within a narrow range of pH and temperature. In the laboratory, the average protein must be boiled for about 24 hours in a 20% HCl solution to achieve a complete breakdown. In the body, the breakdown takes place in four hours or less under conditions of mild physiological temperature and pH.

It is through attempts at understanding more about enzyme catalysts – what they are, what they do, and how they do it– that many advances in medicine and the life sciences have been brought about.

All known enzymes are proteins. They are high molecular weight compounds made up principally of chains of amino acids linked together by peptide bonds.

The primary benefit of enzymes for us is in digestion...

Digestive Enzymes

Specific enzymes work on specific foods. You need the right type of enzyme for the foods you want it to break down. Digestive enzymes are enzymes that break down food into usable material. The major different types of digestive enzymes are:

- amylase – breaks down carbohydrates, starches, and sugars which are prevalent in potatoes, fruits, vegetables, and many snack foods

- protease – breaks down proteins found in meats, nuts, eggs, and cheese
- lipase – breaks down fats found in most dairy products, nuts, oils, and meat
- cellulase – breaks down cellulose, plant fibre; not found in humans

Enzyme is a Biological Catalyst

Catalyst is a molecule that increases the rate of a chemical reaction without itself being permanently changed after the reaction.

Enzymes are proteins that catalyse (*i.e.*, increase or decrease the rates of) chemical reactions. In enzymatic reactions, the molecules at the beginning of the process are called substrates, and they are converted into different molecules, called the products. Almost all processes in a biological cell need enzymes to occur at significant rates. Since enzymes are selective for their substrates and speed up only a few reactions from among many possibilities, the set of enzymes made in a cell determines which metabolic pathways occur in that cell.

Like all catalysts, enzymes work by lowering the activation energy ($E_a^{\ddagger}$) for a reaction, thus dramatically increasing the rate of the reaction. As a result, products are formed faster and reactions reach their equilibrium state more rapidly. Most enzyme reaction rates are millions of times faster than those of comparable un-catalysed reactions. As with all catalysts, enzymes are not consumed by the reactions they catalyse, nor do they alter the equilibrium of these reactions. However, enzymes do differ from most other catalysts by being much more specific. Enzymes are known to catalyse about 4,000 biochemical reactions. A few RNA molecules called ribozymes also catalyse reactions, with an important example being some parts of the ribosome. Synthetic molecules called artificial enzymes also display enzyme-like catalysis.

Enzyme activity can be affected by other molecules. Inhibitors are molecules that decrease enzyme activity; activators are molecules that increase activity. Many drugs and poisons are enzyme inhibitors. Activity is also affected by temperature, chemical environment (*e.g.*, pH), and the concentration of substrate. Some enzymes are used commercially, for example, in the synthesis of antibiotics. In addition, some household products use enzymes to speed up biochemical reactions (*e.g.*, enzymes in biological washing powders break down protein or fat stains on clothes; enzymes in meat tenderizers break down proteins, making the meat easier to chew).

Structures and Mechanisms

Enzymes are generally globular proteins and range from just 62 amino acid residues in size, for the monomer of 4-oxalocrotonate tautomerase, to over

2,500 residues in the animal fatty acid synthase. A small number of RNA-based biological catalysts exist, with the most common being the ribosome; these are referred to as either RNA-enzymes or ribozymes. The activities of enzymes are determined by their three-dimensional structure. However, although structure does determine function, predicting a novel enzyme's activity just from its structure is a very difficult problem that has not yet been solved.

Most enzymes are much larger than the substrates they act on, and only a small portion of the enzyme (around 3–4 amino acids) is directly involved in catalysis. The region that contains these catalytic residues, binds the substrate, and then carries out the reaction is known as the active site.

Enzymes can also contain sites that bind cofactors, which are needed for catalysis. Some enzymes also have binding sites for small molecules, which are often direct or indirect products or substrates of the reaction catalysed. This binding can serve to increase or decrease the enzyme's activity, providing a means for feedback regulation.

Like all proteins, enzymes are long, linear chains of amino acids that fold to produce a three-dimensional product. Each unique amino acid sequence produces a specific structure, which has unique properties. Individual protein chains may sometimes group together to form a protein complex. Most enzymes can be denatured—that is, unfolded and inactivated—by heating or chemical denaturants, which disrupt the three-dimensional structure of the protein. Depending on the enzyme, denaturation may be reversible or irreversible.

Structures of enzymes in complex with substrates or substrate analogs during a reaction may be obtained using Time resolved crystallography methods.

Specificity

Enzymes are usually very specific as to which reactions they catalyze and the substrates that are involved in these reactions. Complementary shape, charge and hydrophilic/hydrophobic characteristics of enzymes and substrates are responsible for this specificity. Enzymes can also show impressive levels of stereospecificity, regioselectivity and chemoselectivity.

Some of the enzymes showing the highest specificity and accuracy are involved in the copying and expression of the genome. These enzymes have "proof-reading" mechanisms. Here, an enzyme such as DNA polymerase catalyzes a reaction in a first step and then checks that the product is correct in a second step. This two-step process results in average error rates of less than 1 error in 100 million reactions in high-fidelity mammalian polymerases. Similar proofreading mechanisms are also found in RNA polymerase, aminoacyl tRNA synthetases and ribosomes.

Some enzymes that produce secondary metabolites are described as promiscuous, as they can act on a relatively broad range of different substrates. It has been suggested that this broad substrate specificity is important for the evolution of new biosynthetic pathways.

"Lock and Key" Model

Enzymes are very specific, and it was suggested by the Nobel laureate organic chemist Emil Fischer in 1894 that this was because both the enzyme and the substrate possess specific complementary geometric shapes that fit exactly into one another. This is often referred to as "the lock and key" model. However, while this model explains enzyme specificity, it fails to explain the stabilization of the transition state that enzymes achieve. In 1958, Daniel Koshland suggested a modification to the lock and key model: since enzymes are rather flexible structures, the active site is continually reshaped by interactions with the substrate as the substrate interacts with the enzyme. As a result, the substrate does not simply bind to a rigid active site; the amino acid side chains which make up the active site are molded into the precise positions that enable the enzyme to perform its catalytic function. In some cases, such as glycosidases, the substrate molecule also changes shape slightly as it enters the active site. The active site continues to change until the substrate is completely bound, at which point the final shape and charge is determined. Induced fit may enhance the fidelity of molecular recognition in the presence of competition and noise via the conformational proofreading mechanism.

Mechanisms

Enzymes can act in several ways, all of which lower $\Delta G^{\ddagger}$:

- Lowering the activation energy by creating an environment in which the transition state is stabilized (e.g. straining the shape of a substrate—by binding the transition-state conformation of the substrate/product molecules, the enzyme distorts the bound substrate(s) into their transition state form, thereby reducing the amount of energy required to complete the transition).
- Lowering the energy of the transition state, but without distorting the substrate, by creating an environment with the opposite charge distribution to that of the transition state.
- Providing an alternative pathway. For example, temporarily reacting with the substrate to form an intermediate ES complex, which would be impossible in the absence of the enzyme.
- Reducing the reaction entropy change by bringing substrates together in the correct orientation to react. Considering $\Delta H^{\ddagger}$ alone overlooks this effect.

- Increases in temperatures speed up reactions. Thus, temperature increases help the enzyme function and develop the end product even faster. However, if heated too much, the enzyme's shape deteriorates and only when the temperature comes back to normal does the enzyme regain its shape. Some enzymes like thermolabile enzymes work best at low temperatures.

Interestingly, this entropic effect involves destabilization of the ground state, and its contribution to catalysis is relatively small.

Transition State Stabilization

The understanding of the origin of the reduction of $\Delta G^{\ddagger}$ requires one to find out how the enzymes can stabilize its transition state more than the transition state of the uncatalyzed reaction. Apparently, the most effective way for reaching large stabilization is the use of electrostatic effects, in particular, by having a relatively fixed polar environment that is oriented towards the charge distribution of the transition state. Such an environment does not exist in the uncatalyzed reaction in water.

Dynamics and Function

The internal dynamics of enzymes is linked to their mechanism of catalysis. Internal dynamics are the movement of parts of the enzyme's structure, such as individual amino acid residues, a group of amino acids, or even an entire protein domain. These movements occur at various time-scales ranging from femtoseconds to seconds.

Networks of protein residues throughout an enzyme's structure can contribute to catalysis through dynamic motions. Protein motions are vital to many enzymes, but whether small and fast vibrations, or larger and slower conformational movements are more important depends on the type of reaction involved. However, although these movements are important in binding and releasing substrates and products, it is not clear if protein movements help to accelerate the chemical steps in enzymatic reactions. These new insights also have implications in understanding allosteric effects and developing new drugs.

Allosteric Modulation

Allosteric sites are sites on the enzyme that bind to molecules in the cellular environment. The sites form weak, noncovalent bonds with these molecules, causing a change in the conformation of the enzyme. This change in conformation translates to the active site, which then affects the reaction rate of the enzyme. Allosteric interactions can both inhibit and activate enzymes and are a common way that enzymes are controlled in the body.

Cofactors and Coenzymes

Cofactors

Some enzymes do not need any additional components to show full activity. However, others require non-protein molecules called cofactors to be bound for activity. Cofactors can be either inorganic (*e.g.*, metal ions and iron-sulphur clusters) or organic compounds (e.g., flavin and heme). Organic cofactors can be either prosthetic groups, which are tightly bound to an enzyme, or coenzymes, which are released from the enzyme's active site during the reaction. Coenzymes include NADH, NADPH and adenosine triphosphate. These molecules transfer chemical groups between enzymes.

An example of an enzyme that contains a cofactor is carbonic anhydrase, and is shown in the ribbon diagram above with a zinc cofactor bound as part of its active site. These tightly bound molecules are usually found in the active site and are involved in catalysis. For example, flavin and heme cofactors are often involved in redox reactions.

Enzymes that require a cofactor but do not have one bound are called *apoenzymes* or *apoproteins*. An apoenzyme together with its cofactor(s) is called a *holoenzyme* (this is the active form). Most cofactors are not covalently attached to an enzyme, but are very tightly bound. However, organic prosthetic groups can be covalently bound (*e.g.*, thiamine pyrophosphate in the enzyme pyruvate dehydrogenase). The term "holoenzyme" can also be applied to enzymes that contain multiple protein subunits, such as the DNA polymerases; here the holoenzyme is the complete complex containing all the subunits needed for activity.

Coenzymes

Coenzymes are small organic molecules that transport chemical groups from one enzyme to another. Some of these chemicals such as riboflavin, thiamine and folic acid are vitamins (compounds which cannot be synthesized by the body and must be acquired from the diet). The chemical groups carried include the hydride ion (H^-) carried by NAD or $NADP^+$, the phosphate group carried by adenosine triphosphate, the acetyl group carried by coenzyme A, formyl, methenyl or methyl groups carried by folic acid and the methyl group carried by S-adenosylmethionine.

Since coenzymes are chemically changed as a consequence of enzyme action, it is useful to consider coenzymes to be a special class of substrates, or second substrates, which are common to many different enzymes. For example, about 700 enzymes are known to use the coenzyme NADH.

Coenzymes are usually continuously regenerated and their concentrations maintained at a steady level inside the cell: for example, NADPH is regenerated through the pentose phosphate pathway and *S*-adenosylmethionine by methionine adenosyltransferase. This continuous regeneration means that even small amounts of coenzymes are used very intensively. For example, the human body turns over its own weight in ATP each day.

Thermodynamics

As all catalysts, enzymes do not alter the position of the chemical equilibrium of the reaction. Usually, in the presence of an enzyme, the reaction runs in the same direction as it would without the enzyme, just more quickly. However, in the absence of the enzyme, other possible uncatalyzed, "spontaneous" reactions might lead to different products, because in those conditions this different product is formed faster. Furthermore, enzymes can couple two or more reactions, so that a thermodynamically favourable reaction can be used to "drive" a thermodynamically unfavourable one. For example, the hydrolysis of ATP is often used to drive other chemical reactions.

Enzymes catalyse the forward and backward reactions equally. They do not alter the equilibrium itself, but only the speed at which it is reached. For example, carbonic anhydrase catalyzes its reaction in either direction depending on the concentration of its reactants.

$$CO_2 + H_2O \xrightarrow{\text{Carbonic anhydrase}} H_2CO_3$$

(in tissues; high CO_2 concentration)

$$H_2CO_3 \xrightarrow{\text{Carbonic anhydrase}} CO_2H_2O$$

(in lungs; low CO_2 concentration)

Nevertheless, if the equilibrium is greatly displaced in one direction, that is, in a very exergonic reaction, the reaction is *effectively* irreversible. Under these conditions the enzyme will, in fact, only catalyse the reaction in the thermodynamically allowed direction.

Kinetics

Enzyme kinetics is the investigation of how enzymes bind substrates and turn them into products. The rate data used in kinetic analyses are obtained from enzyme assays.

In 1902 Victor Henri proposed a quantitative theory of enzyme kinetics, but his experimental data were not useful because the significance of the hydrogen ion concentration was not yet appreciated. After Peter Lauritz Sørensen had defined the logarithmic pH-scale and introduced the concept of buffering. In 1909 the German chemist Leonor Michaelis and his Canadian postdoc Maud Leonora Menten repeated Henri's experiments and confirmed his equation which is referred to as Henri-Michaelis-Menten kinetics (sometimes also Michaelis-Menten kinetics). Their work was further developed by G. E. Briggs and J. B. S. Haldane, who derived kinetic equations that are still widely used today. The major contribution of Henri was to think of enzyme reactions in two stages. In the first, the substrate binds reversibly to the enzyme, forming the enzyme-substrate complex. This is sometimes called the Michaelis complex. The enzyme then catalyzes the chemical step in the reaction and releases the product.

Enzymes can catalyse up to several million reactions per second. For example, the uncatalyzed decarboxylation of orotidine 5'-monophosphate has a half life of 78 million years. However, when the enzyme orotidine 5'-phosphate decarboxylase is added, the same process takes just 25 milliseconds. Enzyme rates depend on solution conditions and substrate concentration. Conditions that denature the protein abolish enzyme activity, such as high temperatures, extremes of pH or high salt concentrations, while raising substrate concentration tends to increase activity. To find the maximum speed of an enzymatic reaction, the substrate concentration is increased until a constant rate of product formation is seen.

This is shown by saturation curve. Saturation happens because, as substrate concentration increases, more and more of the free enzyme is converted into the substrate-bound ES form. At the maximum velocity (V_{max}) of the enzyme, all the enzyme active sites are bound to substrate, and the amount of ES complex is the same as the total amount of enzyme. However, V_{max} is only one kinetic constant of enzymes. The amount of substrate needed to achieve a given rate of reaction is also important. This is given by the Michaelis-Menten constant (K_m), which is the substrate concentration required for an enzyme to reach one-half its maximum velocity. Each enzyme has a characteristic K_m for a given substrate, and this can show how tight the binding of the substrate is to the enzyme. Another useful constant is k_{cat}, which is the number of substrate molecules handled by one active site per second.

The efficiency of an enzyme can be expressed in terms of k_{cat}/K_m. This is also called the specificity constant and incorporates the rate constants for all steps in the reaction. Because the specificity constant reflects both affinity and catalytic ability, it is useful for comparing different enzymes against each other, or the same enzyme with different substrates. The theoretical maximum for the specificity constant is called the diffusion limit and is about 10^8 to 10^9 (M^{-1} s^{-1}). At this point every collision of the enzyme with its substrate will result in catalysis, and the rate of product formation is not limited by the reaction rate but by the diffusion rate. Enzymes with this property are called *catalytically perfect* or *kinetically perfect*. Example of such enzymes are triose-phosphate isomerase, carbonic anhydrase, acetylcholinesterase, catalase, fumarase, β-lactamase, and superoxide dismutase.

Michaelis-Menten kinetics relies on the law of mass action, which is derived from the assumptions of free diffusion and thermodynamically driven random collision. However, many biochemical or cellular processes deviate significantly from these conditions, because of macromolecular crowding, phase-separation of the enzyme/substrate/product, or one or two-dimensional molecular movement. In these situations, a fractal Michaelis-Menten kinetics may be applied.

Some enzymes operate with kinetics which are faster than diffusion rates, which would seem to be impossible. Several mechanisms have been invoked to explain this phenomenon. Some proteins are believed to accelerate catalysis by drawing their substrate in and pre-orienting them by using dipolar electric fields. Other models invoke a quantum-mechanical tunnelling explanation, whereby a proton or an electron can tunnel through activation barriers, although for proton tunnelling this model remains somewhat controversial. Quantum tunnelling for protons has been observed in tryptamine. This suggests that enzyme catalysis may be more accurately characterized as "through the barrier" rather than the traditional model, which requires substrates to go "over" a lowered energy barrier.

Inhibition

Enzyme inhibitors are molecules that bind to enzymes and decrease their activity. Since blocking an enzyme's activity can kill a pathogen or correct a metabolic imbalance, many drugs are enzyme inhibitors. They are also used as herbicides and pesticides. Not all molecules that bind to enzymes are inhibitors; *enzyme activators* bind to enzymes and increase their enzymatic activity.

The binding of an inhibitor can stop a substrate from entering the enzyme's active site and/or hinder the enzyme from catalysing its reaction. Inhibitor binding is either reversible or irreversible. Irreversible inhibitors usually react with the enzyme and change it chemically. These inhibitors modify key amino acid residues needed for enzymatic activity. In contrast, reversible inhibitors bind non-covalently and different types of inhibition are produced depending on whether these inhibitors bind the enzyme, the enzyme-substrate complex, or both.

Many drug molecules are enzyme inhibitors, so their discovery and improvement is an active area of research in biochemistry and pharmacology. A medicinal enzyme inhibitor is often judged by its specificity (its lack of binding to other proteins) and its potency (its dissociation constant, which indicates the concentration needed to inhibit the enzyme). A high specificity and potency ensure that a drug will have few side effects and thus low toxicity. Enzyme inhibitors also occur naturally and are involved in the regulation of metabolism. For example, enzymes in a metabolic pathway can be inhibited by downstream products. This type of negative feedback slows flux through a pathway when the products begin to build up and is an important way to maintain homeostasis in a cell. Other cellular enzyme inhibitors are proteins that specifically bind to and inhibit an enzyme target. This can help control enzymes that may be damaging to a cell, such as proteases or nucleases; a well-characterised example is the ribonuclease inhibitor, which binds to ribonucleases in one of the tightest known protein–protein interactions. Natural enzyme inhibitors can also be poisons and are used as defences against predators or as ways of killing prey.

Reversible Inhibitors

Types of Reversible Inhibitors

Reversible inhibitors bind to enzymes with non-covalent interactions such as hydrogen bonds, hydrophobic interactions and ionic bonds. Multiple weak bonds between the inhibitor and the active site combine to produce strong and specific binding. In contrast to substrates and irreversible inhibitors, reversible inhibitors generally do not undergo chemical reactions when bound to the enzyme and can be easily removed by dilution or dialysis.

There are four kinds of reversible enzyme inhibitors. They are classified according to the effect of varying the concentration of the enzyme's substrate on the inhibitor.

- In competitive inhibition, the substrate and inhibitor cannot bind to the enzyme at the same time, as shown in the figure on the left. This usually results from the inhibitor having an affinity for the active site of an enzyme where the substrate also binds; the substrate and inhibitor *compete* for access to the enzyme's active site. This type of inhibition can be overcome by sufficiently high concentrations of substrate, i.e., by out-competing the inhibitor. Competitive inhibitors are often similar in structure to the real substrate.
- In uncompetitive inhibition, the inhibitor binds only to the substrate-enzyme complex, it should not be confused with non-competitive inhibitors. Both Vmax and Km decrease (maximum velocity decreases as a result of removing activated complex while binding efficiency increases as a result of Le Chatelier's principle).
- In mixed inhibition, the inhibitor can bind to the enzyme at the same time as the enzyme's substrate. However, the binding of the inhibitor affects the binding of the substrate, and vice versa. This type of inhibition can be reduced, but not overcome by increasing concentrations of substrate. Although it is possible for mixed-type inhibitors to bind in

the active site, this type of inhibition generally results from an allosteric effect where the inhibitor binds to a different site on an enzyme. Inhibitor binding to this allosteric site changes the conformation (i.e., tertiary structure or three-dimensional shape) of the enzyme so that the affinity of the substrate for the active site is reduced.

- Non-competitive inhibition is a form of mixed inhibition where the binding of the inhibitor to the enzyme reduces its activity but does not affect the binding of substrate. As a result, the extent of inhibition depends only on the concentration of the inhibitor.

Quantitative Description of Reversible Inhibition

Reversible inhibition can be described quantitatively in terms of the inhibitor's binding to the enzyme and to the enzyme–substrate complex, and its effects on the kinetic constants of the enzyme. In the classic Michaelis–Menten scheme below, an enzyme (E) binds to its substrate (S) to form the enzyme–substrate complex ES. Upon catalysis, this complex breaks down to release product P and free enzyme. The inhibitor (I) can bind to either E or ES with the dissociation constants K_i or K_i', respectively.

When an enzyme has multiple substrates, inhibitors can show different types of inhibition depending on which substrate is considered. This results from the active site containing two different binding sites within the active site, one for each substrate. For example, an inhibitor might compete with substrate A for the first binding site, but be a non-competitive inhibitor with respect to substrate B in the second binding site.

Measuring the Dissociation Constants of a Reversible Inhibitor

As noted above, an enzyme inhibitor is characterised by its two dissociation constants, K_i and K_i', to the enzyme and to the enzyme-substrate complex, respectively. The enzyme-inhibitor constant K_i can be measured directly by various methods; one extremely accurate method is isothermal titration calorimetry, in which the inhibitor is titrated into a solution of enzyme and the heat released or absorbed is measured. However, the other dissociation constant K_i' is difficult to measure directly, since the enzyme-substrate complex is short-lived and undergoing a chemical reaction to form the product. Hence, K_i' is usually measured indirectly, by observing the enzyme activity under various substrate and inhibitor concentrations, and fitting the data to a modified Michaelis–Menten equation

$$V = \frac{V_{max}[S]}{\alpha K_m + \alpha'[S]} = \frac{(1/\alpha')V_{max}[S]}{(\alpha/\alpha')K_m + [S]}$$

where the modifying factors α and α' are defined by the inhibitor concentration and its two dissociation constants

$$\alpha = 1 + \frac{[1]}{K_i}$$

$$\alpha' = 1 + \frac{[I]}{K_i'}.$$

Thus, in the presence of the inhibitor, the enzyme's effective K_m and V_{max} become $(\alpha/\alpha')K_m$ and $(1/\alpha')V_{max}$, respectively. However, the modified Michaelis-Menten equation assumes that binding of the inhibitor to the enzyme has reached equilibrium, which may be a very slow process for inhibitors with sub-nanomolar dissociation constants.

In these cases, it is usually more practical to treat the tight-binding inhibitor as an irreversible inhibitor; however, it can still be possible to estimate K_i' kinetically if K_i is measured independently.

The effects of different types of reversible enzyme inhibitors on enzymatic activity can be visualized using graphical representations of the Michaelis–Menten equation, such as Lineweaver–Burk and Eadie-Hofstee plots. For example, in the Lineweaver–Burk plots at the right, the competitive inhibition lines intersect on the y-axis, illustrating that such inhibitors do not affect V_{max}. Similarly, the non-competitive inhibition lines intersect on the x-axis, showing these inhibitors do not affect K_m. However, it can be difficult to estimate K_i and K_i' accurately from such plots, so it is advisable to estimate these constants using more reliable nonlinear regression methods, as described above.

Special Cases

- The mechanism of partially competitive inhibition is similar to that of non-competitive, except that the EIS complex has catalytic activity, which may be lower or even higher (partially competitive activation) than that of the enzyme–substrate (ES) complex. This inhibition typically displays a lower V_{max}, but an unaffected K_m value.
- Uncompetitive inhibition occurs when the inhibitor binds only to the enzyme–substrate complex, not to the free enzyme; the EIS complex is catalytically inactive. This mode of inhibition is rare and causes a decrease in both V_{max} and the K_m value.
- Substrate and product inhibition is where either the substrate or product of an enzyme reaction inhibit the enzyme's activity. This inhibition may follow the competitive, uncompetitive or mixed patterns. In substrate inhibition there is a progressive decrease in activity at high substrate concentrations. This may indicate the existence of two substrate-binding sites in the enzyme. At low substrate, the high-affinity site is occupied and normal kinetics are followed. However, at higher concentrations,

the second inhibitory site becomes occupied, inhibiting the enzyme. Product inhibition is often a regulatory feature in metabolism and can be a form of negative feedback.

- Slow-tight inhibition occurs when the initial enzyme–inhibitor complex EI undergoes isomerisation to a second more tightly held complex, EI*, but the overall inhibition process is reversible. This manifests itself as slowly increasing enzyme inhibition. Under these conditions, traditional Michaelis–Menten kinetics give a false value for K_i, which is time–dependent. The true value of K_i can be obtained through more complex analysis of the on (k_{on}) and off (k_{off}) rate constants for inhibitor association.

Examples of Reversible Inhibitors

As enzymes have evolved to bind their substrates tightly, and most reversible inhibitors bind in the active site of enzymes, it is unsurprising that some of these inhibitors are strikingly similar in structure to the substrates of their targets. An example of these substrate mimics are the protease inhibitors, a very successful class of antiretroviral drugs used to treat HIV. The structure of ritonavir, a protease inhibitor based on a peptide and containing three peptide bonds, is shown on the right. As this drug resembles the protein that is the substrate of the HIV protease, it competes with this substrate in the enzyme's active site.

Enzyme inhibitors are often designed to mimic the transition state or intermediate of an enzyme-catalysed reaction. This ensures that the inhibitor exploits the transition state stabilising effect of the enzyme, resulting in a better binding affinity (lower K_i) than substrate-based designs. An example of such a transition state inhibitor is the antiviral drug oseltamivir; this drug mimics the planar nature of the ring oxonium ion in the reaction of the viral enzyme neuraminidase.

However, not all inhibitors are based on the structures of substrates. For example, the structure of another HIV protease inhibitor tipranavir is shown on the left. This molecule is not based on a peptide and has no obvious structural similarity to a protein substrate. These non-peptide inhibitors can be more stable than inhibitors containing peptide bonds, because they will not be substrates for peptidases and are less likely to be degraded.

In drug design it is important to consider the concentrations of substrates to which the target enzymes are exposed. For example, some protein kinase inhibitors have chemical structures that are similar to adenosine triphosphate, one of the substrates of these enzymes. However, drugs that are simple competitive inhibitors will have to compete with the high concentrations of ATP in the cell. Protein kinases can also be inhibited by competition at the binding sites where the kinases interact with their substrate proteins, and most proteins are present inside cells at concentrations much lower than the concentration of ATP. As a consequence, if two protein kinase inhibitors both bind in the active site with similar affinity, but only one has to compete with ATP, then the competitive inhibitor at the protein-binding site will inhibit the enzyme more effectively.

Irreversible Inhibitors

Types of Irreversible Inhibition

Irreversible inhibitors usually covalently modify an enzyme, and inhibition cannot therefore be reversed. Irreversible inhibitors often contain reactive functional groups such as nitrogen mustards, aldehydes, haloalkanes, alkenes, Michael acceptors, phenyl sulphonates, or fluorophosphonates. These electrophilic groups react with amino acid side chains to form covalent adducts. The residues modified are those with side chains containing nucleophiles such as hydroxyl or sulphhydryl groups; these include the amino acids serine (as in DFP, right), cysteine, threonine or tyrosine.

Irreversible inhibition is different from irreversible enzyme inactivation. Irreversible inhibitors are generally specific for one class of enzyme and do not inactivate all proteins; they do not function by destroying protein structure but by specifically altering the active site of their target. For example, extremes of pH or temperature usually cause denaturation of all protein structure, but this is a non-specific effect. Similarly, some non-specific chemical treatments destroy protein structure: for example, heating in concentrated hydrochloric acid will hydrolyse the peptide bonds holding proteins together, releasing free amino acids.

Irreversible inhibitors display time-dependent inhibition and their potency therefore cannot be characterized by an IC_{50} value. This is because the amount of active enzyme at a given concentration of irreversible inhibitor will be different depending on how long the inhibitor is pre-incubated with the enzyme. Instead, $k_{obs}/[I]$ values are used, where k_{obs} is the observed pseudo-first order rate of inactivation (obtained by plotting the log of % activity vs time) and $[I]$ is the concentration of inhibitor. The $k_{obs}/[I]$ parameter is valid as long as the inhibitor does not saturate binding with the enzyme (in which case $k_{obs} = k_{inact}$).

Analysis of Irreversible Inhibition

The irreversible inhibitors form a reversible non-covalent complex with the enzyme (EI or ESI) and this then reacts to produce the covalently modified "dead-end complex" EI*. The rate at which EI* is formed is called the inactivation rate or k_{inact}. Since formation of EI may compete with ES, binding of irreversible inhibitors can be prevented by competition either with substrate or with a second, reversible inhibitor. This protection effect is good evidence of a specific reaction of the irreversible inhibitor with the active site.

The binding and inactivation steps of this reaction are investigated by incubating the enzyme with inhibitor and assaying the amount of activity remaining over time. The activity will be decrease in a time-dependent manner, usually following exponential decay. Fitting these data to a rate equation gives the rate of inactivation at this concentration of inhibitor. This is done at several different concentrations of inhibitor. If a reversible EI complex is involved the inactivation rate will be saturable and fitting this curve will give k_{inact} and K_i. Another method that is widely used in these analyses is mass spectrometry. Here, accurate measurement of the mass of the unmodified native enzyme and the inactivated enzyme gives the increase in mass caused by reaction with the inhibitor and shows the stoichiometry of the reaction. This is usually done using a MALDI-TOF mass spectrometer. In a complementary technique, peptide mass fingerprinting involves digestion of the native and modified protein with a protease such as trypsin. This will produce a set of peptides that can be analysed using a mass spectrometer. The peptide that changes in mass after reaction with the inhibitor will be the one that contains the site of modification.

Special Cases

Not all irreversible inhibitors form covalent adducts with their enzyme targets. Some reversible inhibitors bind so tightly to their target enzyme that they are essentially irreversible. These tight-binding inhibitors may show kinetics similar to covalent irreversible inhibitors. In these cases, some of these inhibitors rapidly bind to the enzyme in a low-affinity EI complex and this then undergoes a slower rearrangement to a very tightly bound EI* complex. This kinetic behaviour is called slow-binding. This slow rearrangement after binding often involves a conformational change as the enzyme "clamps down" around the inhibitor molecule. Examples of slow-binding inhibitors include some important drugs, such methotrexate, allopurinol, and the activated form of acyclovir.

Examples of Irreversible Inhibitors

Diisopropylfluorophosphate (DFP) is shown as an example of an irreversible protease inhibitor in the figure above right. The enzyme hydrolyses the phosphorus–fluorine bond, but the phosphate residue remains bound to the serine in the active site, deactivating it. Similarly, DFP also reacts with the active site of acetylcholine esterase in the synapses of neurons, and consequently is a potent neurotoxin, with a lethal dose of less than 100 mg. Suicide inhibition is an unusual type of irreversible inhibition where the enzyme converts the inhibitor into a reactive form in its active site. An example is the inhibitor of polyamine biosynthesis, α-difluoromethylornithine or DFMO, which is an analogue of the amino acid ornithine, and is used to treat African trypanosomiasis (sleeping sickness). Ornithine decarboxylase can catalyse the decarboxylation of DFMO instead of ornithine, as shown above. However, this decarboxylation reaction is followed by the elimination of a fluorine atom, which converts this catalytic intermediate into a conjugated imine, a highly electrophilic species. This reactive form of DFMO then reacts with either a cysteine or lysine residue in the active site to irreversibly inactivate the enzyme.

Since irreversible inhibition often involves the initial formation of a non-covalent EI complex, it is sometimes possible for an inhibitor to bind to an enzyme in more than one way. For example, in the trypanothione reductase from the human protozoan parasite *Trypanosoma cruzi*, two molecules of an inhibitor called *quinacrine mustard* are bound in its active site. The top molecule is bound reversibly, but the lower one is bound covalently as it has reacted with an amino acid residue through its nitrogen mustard group.

Discovery and Design of Inhibitors

New drugs are the products of a long drug development process, the first step of which is often the discovery of a new enzyme inhibitor. In the past the only way to discover these new inhibitors was by trial and error: screening huge libraries of compounds against a target enzyme and hoping that some useful leads would emerge. This brute force approach is still successful and has even been extended by combinatorial chemistry approaches that quickly produce large numbers of novel compounds and high-throughput screening technology to rapidly screen these huge chemical libraries for useful inhibitors.

More recently, an alternative approach has been applied: rational drug design uses the three-dimensional structure of an enzyme's active site to predict which molecules might be inhibitors. These predictions are then tested and one of these tested compounds may be a novel inhibitor. This new inhibitor is then used to try to obtain a structure of the enzyme in an inhibitor/enzyme complex to show how the molecule is binding to the active site, allowing changes to be made to the inhibitor to try to optimise binding. This test and improve cycle is then repeated until a sufficiently potent inhibitor is produced. Computer-based methods of predicting the affinity of an inhibitor for an enzyme are also being developed, such as molecular docking and molecular mechanics.

Uses of Inhibitors

Enzyme inhibitors are found in nature and are also designed and produced as part of pharmacology and biochemistry. Natural poisons are often enzyme inhibitors that have evolved to defend a plant or animal against predators.

These natural toxins include some of the most poisonous compounds known. Artificial inhibitors are often used as drugs, but can also be insecticides such as malathion, herbicides such as glyphosate, or disinfectants such as triclosan.

Chemotherapy

The most common uses for enzyme inhibitors are as drugs to treat disease. Many of these inhibitors target a human enzyme and aim to correct a pathological condition. However, not all drugs are enzyme inhibitors. Some, such as anti-epileptic drugs, alter enzyme activity by causing more or less of the enzyme to be produced. These effects are called enzyme induction and inhibition and are alterations in gene expression, which is unrelated to the type of enzyme inhibition.

Other drugs interact with cellular targets that are not enzymes, such as ion channels or membrane receptors. An example of a medicinal enzyme inhibitor is sildenafil (Viagra), a common treatment for male erectile dysfunction. This compound is a potent inhibitor of cGMP specific phosphodiesterase type 5, the enzyme that degrades the signalling molecule cyclic guanosine monophosphate.

This signalling molecule triggers smooth muscle relaxation and allows blood flow into the corpus cavernosum, which causes an erection. Since the drug decreases the activity of the enzyme that halts the signal, it makes this signal last for a longer period of time.

Another example of the structural similarity of some inhibitors to the substrates of the enzymes they target is seen in the figure comparing the drug methotrexate to folic acid. Folic acid is a substrate of dihydrofolate reductase, an enzyme involved in making nucleotides that is potently inhibited by methotrexate. Methotrexate blocks the action of dihydrofolate reductase and thereby halts the production of nucleotides. This block of nucleotide biosynthesis is more toxic to rapidly growing cells than non-dividing cells, since a rapidly-growing cell has to carry out DNA replication, therefore methotrexate is often used in cancer chemotherapy.

Drugs also are used to inhibit enzymes needed for the survival of pathogens. For example, bacteria are surrounded by a thick cell wall made of a net-like polymer called peptidoglycan. Many antibiotics such as penicillin and vancomycin inhibit the enzymes that produce and then cross-link the strands of this polymer together. This causes the cell wall to lose strength and the bacteria to burst.

Drug design is facilitated when an enzyme that is essential to the pathogen's survival is absent or very different in humans. In the example above, humans do not make peptidoglycan, therefore inhibitors of this process are selectively toxic to bacteria. Selective toxicity is also produced in antibiotics by exploiting differences in the structure of the ribosomes in bacteria, or how they make fatty acids.

Metabolic Control

Enzyme inhibitors are also important in metabolic control. Many metabolic pathways in the cell are inhibited by metabolites that control enzyme activity through allosteric regulation or substrate inhibition. A good example is the allosteric regulation of the glycolytic pathway. This catabolic pathway consumes glucose and produces ATP, NADH and pyruvate. A key step for the regulation of glycolysis is an early reaction in the pathway catalysed by phosphofructokinase-1 (PFK1). When ATP levels rise, ATP binds an allosteric site in PFK1 to decrease the rate of the enzyme reaction; glycolysis is inhibited and ATP production falls. This negative feedback control helps maintain a steady concentration of ATP in the cell. However, metabolic pathways are not just regulated through inhibition since enzyme activation is equally important. With respect to PFK1, fructose 2,6-bisphosphate and ADP are examples of metabolites that are allosteric activators.

Physiological enzyme inhibition can also be produced by specific protein inhibitors. This mechanism occurs in the pancreas, which synthesises many digestive precursor enzymes known as zymogens. Many of these are activated by the trypsin protease, so it is important to inhibit the activity of trypsin in the pancreas to prevent the organ from digesting itself. One way in which the activity of trypsin is controlled is the production of a specific and potent trypsin inhibitor protein in the pancreas. This inhibitor binds tightly to trypsin, preventing the trypsin activity that would otherwise be detrimental to the organ. Although the trypsin inhibitor is a protein, it avoids being hydrolysed as a substrate by the protease by excluding water from trypsin's active site and destabilising the transition state. Other examples of physiological enzyme inhibitor proteins include the barstar inhibitor of the bacterial ribonuclease barnase and the inhibitors of protein phosphatases.

Pesticides and Herbicides

Many herbicides and pesticides are enzyme inhibitors. Acetylcholinesterase (AChE) is an enzyme found in animals from insects to humans. It is essential to nerve cell function through its mechanism of breaking down the neurotransmitter acetylcholine into its constituents, acetate and choline. This is somewhat unique among neurotransmitters as most, including serotonin, dopamine, and norepinephrine, are absorbed from the synaptic cleft rather than cleaved. A large number of AChE inhibitors are used in both medicine and agriculture. Reversible competitive inhibitors, such as edrophonium, physostigmine, and neostigmine, are used in the treatment of myasthenia gravis and in anaesthesia. The carbamate pesticides are also examples of reversible AChE inhibitors. The organophosphate insecticides such as malathion, parathion, and chlorpyrifos irreversibly inhibit acetylcholinesterase.

The herbicide glyphosate is an inhibitor of 3-phosphoshikimate 1-carboxyvinyltransferase, other herbicides, such as the sulphonylureas inhibit the enzyme acetolactate synthase. Both these enzymes are needed for plants to make branched-chain amino acids. Many other enzymess are inhibited by herbicides, including enzymes needed for the biosynthesis of lipids and carotenoids and the processes of photosynthesis and oxidative phosphorylation.

Natural Poisons

Animals and plants have evolved to synthesise a vast array of poisonous products including secondary metabolites, peptides and proteins that can act as inhibitors. Natural toxins are usually small organic molecules and are so diverse that there are probably natural inhibitors for most metabolic processes. The metabolic processes targeted by natural poisons encompass more than enzymes in metabolic pathways and can also include the inhibition of receptor, channel and structural protein functions in a cell. For example, paclitaxel (taxol), an organic molecule found in the Pacific yew tree, binds tightly to tubulin dimers and inhibits their assembly into microtubules in the cytoskeleton.

Many natural poisons act as neurotoxins that can cause paralysis leading to death and have functions for defence against predators or in hunting and capturing prey. Some of these natural inhibitors, despite their toxic attributes, are valuable for therapeutic uses at lower doses. An example of a neurotoxin are the glycoalkaloids, from the plant species in the *Solanaceae* family (includes potato, tomato and eggplant), that are acetylcholinesterase inhibitors. Inhibition of this enzyme causes an uncontrolled increase in the acetylcholine neurotransmitter, muscular paralysis and then death. Neurotoxicity can also result from the inhibition of receptors; for example, atropine from deadly nightshade (*Atropa belladonna*) that functions as a competitive antagonist of the muscarinic acetylcholine receptors.

Although many natural toxins are secondary metabolites, these poisons also include peptides and proteins. An example of a toxic peptide is alpha-amanitin, which is found in relatives of the death cap mushroom. This is a potent enzyme inhibitor, in this case preventing the RNA polymerase II enzyme from transcribing DNA. The algal toxin microcystin is also a peptide and is an inhibitor of protein phosphatases. This toxin can contaminate water supplies after algal blooms and is a known carcinogen that can also cause acute liver hemorrhage and death at higher doses. Proteins can also be natural poisons or antinutrients, such as the trypsin inhibitors that are found in some legumes. A less common class of toxins are toxic enzymes: these act as irreversible inhibitors of their target enzymes and work by chemically modifying their substrate enzymes. An example is ricin, an extremely potent protein toxin found in castor oil beans. This enzyme is a glycosidase that inactivates ribosomes. Since ricin is a catalytic irreversible inhibitor, this allows just a single molecule of ricin to kill a cell. Enzyme reaction rates can be decreased by various types of enzyme inhibitors.

Competitive Inhibition

In competitive inhibition, the inhibitor and substrate compete for the enzyme (i.e., they can not bind at the same time). Often competitive inhibitors strongly resemble the real substrate of the enzyme. For example, methotrexate is a competitive inhibitor of the enzyme dihydrofolate reductase, which catalyzes the reduction of dihydrofolate to tetrahydrofolate.

The similarity between the structures of folic acid and this drug. Note that binding of the inhibitor need *not* be to the substrate binding site (as frequently stated), if binding of the inhibitor changes the conformation of the enzyme to prevent substrate binding and *vice versa*. In competitive inhibition the maximal velocity of the reaction is not changed, but higher substrate concentrations are required to reach a given velocity, increasing the apparent K_m.

Uncompetitive Inhibition

In uncompetitive inhibition the inhibitor can not bind to the free enzyme, but only to the ES-complex. The EIS-complex thus formed is enzymatically inactive. This type of inhibition is rare, but may occur in multimeric enzymes.

Non-competitive Inhibition

Non-competitive inhibitors can bind to the enzyme at the same time as the substrate, i.e. they *never* bind to the active site. Both the EI and EIS complexes are enzymatically inactive. Because the inhibitor can not be driven from the enzyme by higher substrate concentration (in contrast to competitive inhibition), the apparent V_{max} changes. But because the substrate can still bind to the enzyme, the K_m stays the same.

Mixed Inhibition

This type of inhibition resembles the non-competitive, except that the EIS-complex has residual enzymatic activity. In many organisms inhibitors may act as part of a feedback mechanism. If an enzyme produces too much of one substance in the organism, that substance may act as an inhibitor for the enzyme at the beginning of the pathway that produces it, causing production of the substance to slow down or stop when there is sufficient amount. This is a form of negative feedback. Enzymes which are subject to this form of regulation are often multimeric and have allosteric binding sites for regulatory substances. Their substrate/velocity plots are not hyperbolar, but sigmoidal

(S-shaped). Irreversible inhibitors react with the enzyme and form a covalent adduct with the protein. The inactivation is irreversible. These compounds include eflornithine a drug used to treat the parasitic disease sleeping sickness. Penicillin and Aspirin also act in this manner. With these drugs, the compound is bound in the active site and the enzyme then converts the inhibitor into an activated form that reacts irreversibly with one or more amino acid residues.

Uses of Inhibitors

Since inhibitors modulate the function of enzymes they are often used as drugs. An common example of an inhibitor that is used as a drug is aspirin, which inhibits the COX-1 and COX-2 enzymes that produce the inflammation messenger prostaglandin, thus suppressing pain and inflammation. However, other enzyme inhibitors are poisons. For example, the poison cyanide is an irreversible enzyme inhibitor that combines with the copper and iron in the active site of the enzyme cytochrome c oxidase and blocks cellular respiration.

Biological Function

Enzymes serve a wide variety of functions inside living organisms. They are indispensable for signal transduction and cell regulation, often via kinases and phosphatases. They also generate movement, with myosin hydrolysing ATP to generate muscle contraction and also moving cargo around the cell as part of the cytoskeleton. Other ATPases in the cell membrane are ion pumps involved in active transport. Enzymes are also involved in more exotic functions, such as luciferase generating light in fireflies. Viruses can also contain enzymes for infecting cells, such as the HIV integrase and reverse transcriptase, or for viral release from cells, like the influenza virus neuraminidase.

An important function of enzymes is in the digestive systems of animals. Enzymes such as amylases and proteases break down large molecules (starch or proteins, respectively) into smaller ones, so they can be absorbed by the intestines. Starch molecules, for example, are too large to be absorbed from the intestine, but enzymes hydrolyse the starch chains into smaller molecules such as maltose and eventually glucose, which can then be absorbed. Different enzymes digest different food substances. In ruminants which have herbivorous diets, microorganisms in the gut produce another enzyme, cellulase to break down the cellulose cell walls of plant fibre.

Several enzymes can work together in a specific order, creating metabolic pathways. In a metabolic pathway, one enzyme takes the product of another enzyme as a substrate. After the catalytic reaction, the product is then passed on to another enzyme. Sometimes more than one enzyme can catalyse the same reaction in parallel, this can allow more complex regulation: with for example a low constant activity being provided by one enzyme but an inducible high activity from a second enzyme. Enzymes determine what steps occur in these pathways. Without enzymes, metabolism would neither progress through the same steps, nor be fast enough to serve the needs of the cell. Indeed, a metabolic pathway such as glycolysis could not exist independently of enzymes. Glucose, for example, can react directly with ATP to become phosphorylated at one or more of its carbons.

In the absence of enzymes, this occurs so slowly as to be insignificant. However, if hexokinase is added, these slow reactions continue to take place except that phosphorylation at carbon 6 occurs so rapidly that if the mixture is tested a short time later, glucose-6-phosphate is found to be the only significant product. Consequently, the network of metabolic pathways within each cell depends on the set of functional enzymes that are present.

Control of Activity

There are five main ways that enzyme activity is controlled in the cell.

1. Enzyme production (transcription and translation of enzyme genes) can be enhanced or diminished by a cell in response to changes in the cell's environment. This form of gene regulation is called enzyme induction and inhibition. For example, bacteria may become resistant to antibiotics such as penicillin because enzymes called beta-lactamases are induced that hydrolyse the crucial beta-lactam ring within the penicillin molecule. Another example are enzymes in the liver called cytochrome P450 oxidases, which are important in drug metabolism. Induction or inhibition of these enzymes can cause drug interactions.
2. Enzymes can be compartmentalized, with different metabolic pathways occurring in different cellular compartments. For example, fatty acids are synthesized by one set of enzymes in the cytosol, endoplasmic reticulum and the Golgi apparatus and used by a different set of enzymes as a source of energy in the mitochondrion, through b-oxidation.
3. Enzymes can be regulated by inhibitors and activators. For example, the end product(s) of a metabolic pathway are often inhibitors for one of the first enzymes of the pathway (usually the first irreversible step, called *committed step*), thus regulating the amount of end product made by the pathways. Such a regulatory mechanism is called a negative feedback mechanism, because the amount of the end product produced is regulated by its own concentration. Negative feedback mechanism can effectively adjust the rate of synthesis of intermediate metabolites according to the demands of the cells. This helps allocate materials and energy economically, and prevents the manufacture of excess end

products. The control of enzymatic action helps to maintain a stable internal environment in living organisms.

4. Enzymes can be regulated through post-translational modification. This can include phosphorylation, myristoylation and glycosylation. For example, in the response to insulin, the phosphorylation of multiple enzymes, including glycogen synthase, helps control the synthesis or degradation of glycogen and allows the cell to respond to changes in blood sugar. Another example of post-translational modification is the cleavage of the polypeptide chain. Chymotrypsin, a digestive protease, is produced in inactive form as chymotrypsinogen in the pancreas and transported in this form to the stomach where it is activated. This stops the enzyme from digesting the pancreas or other tissues before it enters the gut. This type of inactive precursor to an enzyme is known as a zymogen.
5. Some enzymes may become activated when localized to a different environment (e.g. from a reducing (cytoplasm) to an oxidizing (periplasm) environment, high pH to low pH etc.). For example, hemagglutinin in the influenza virus is activated by a conformational change caused by the acidic conditions, these occur when it is taken up inside its host cell and enters the lysosome.

Involvement in Disease

Since the tight control of enzyme activity is essential for homeostasis, any malfunction (mutation, overproduction, underproduction or deletion) of a single critical enzyme can lead to a genetic disease. The importance of enzymes is shown by the fact that a lethal illness can be caused by the malfunction of just one type of enzyme out of the thousands of types present in our bodies.

One example is the most common type of phenylketonuria. A mutation of a single amino acid in the enzyme phenylalanine hydroxylase, which catalyzes the first step in the degradation of phenylalanine, results in build-up of phenylalanine and related products. This can lead to mental retardation if the disease is untreated. Another example is when germline mutations in genes coding for DNA repair enzymes cause hereditary cancer syndromes such as xeroderma pigmentosum. Defects in these enzymes cause cancer since the body is less able to repair mutations in the genome. This causes a slow accumulation of mutations and results in the development of many types of cancer in the sufferer.

Naming Conventions

An enzyme's name is often derived from its substrate or the chemical reaction it catalyzes, with the word ending in *-ase*. Examples are lactase, alcohol dehydrogenase and DNA polymerase. This may result in different enzymes, called isozymes, with the same function having the same basic name.

Isoenzymes have a different amino acid sequence and might be distinguished by their optimal pH, kinetic properties or immunologically. Furthermore, the normal physiological reaction an enzyme catalyzes may not be the same as under artificial conditions. This can result in the same enzyme being identified with two different names. *E.g.* Glucose isomerase, used industrially to convert glucose into the sweetener fructose, is a xylose isomerase *in vivo*.

The International Union of Biochemistry and Molecular Biology have developed a nomenclature for enzymes, the EC numbers; each enzyme is described by a sequence of four numbers preceded by "EC". The first number broadly classifies the enzyme based on its mechanism.

The top-level classification is:

- EC 1 *Oxidoreductases*: catalyse oxidation/reduction reactions
- EC 2 *Transferases*: transfer a functional group (*e.g.* a methyl or phosphate group)
- EC 3 *Hydrolases*: catalyse the hydrolysis of various bonds
- EC 4 *Lyases*: cleave various bonds by means other than hydrolysis and oxidation
- EC 5 *Isomerases*: catalyse isomerization changes within a single molecule
- EC 6 *Ligases*: join two molecules with covalent bonds.

According to the naming conventions, enzymes are generally classified into six main family classes and many sub-family classes. Some web-servers, e.g., EzyPred and bioinformatics tools have been developed to predict which main family class and sub-family class an enzyme molecule belongs to according to its sequence information alone via the pseudo amino acid composition.

Industrial Applications

Enzymes are used in the chemical industry and other industrial applications when extremely specific catalysts are required. However, enzymes in general are limited in the number of reactions they have evolved to catalyse and also by their lack of stability in organic solvents and at high temperatures. Consequently, protein engineering is an active area of research and involves attempts to create new enzymes with novel properties, either through rational design or *in vitro* evolution. These efforts have begun to be successful, and a few enzymes have now been desiged "from scratch" to catalyse reactions that do not occur in nature.

6

Understanding DNA and RNA

Nucleic Acid

Nucleic acids are biological molecules essential for life, and include DNA (deoxyribonucleic acid) and RNA (ribonucleic acid). Together with proteins, nucleic acids make up the most important macromolecules; each is found in abundance in all living things. Nucleic acids were first discovered by Friedrich Miescher in 1871. Experimental studies of nucleic acids constitute a major part of modern biological and medical research, and form a foundation for genome and forensic science, as well as the biotechnology and pharmaceutical industries.

Occurrence and Nomenclature

The term *nucleic acid* is the over all name for DNA and RNA, members of a family of biopolymers, and is synonymous with *polynucleotide*. Nucleic acids were named for their initial discovery within the cell nucleus, and for the presence of phosphate groups (related to phosphoric acid).

Although first discovered within the nucleus of eukaryotic cells, nucleic acids are now known to be found in all life forms, including within bacteria, archaea, mitochondria, chloroplasts, viruses and viroids. All living cells and organelles contain both DNA and RNA, while viruses contain either DNA or RNA, but not usually both.

The basic component of biological nucleic acids is the nucleotide, each of which contains a pentose sugar (ribose or deoxyribose), a phosphate group, and a nucleobase. Nucleic acids are also generated within the laboratory, through the use of enzymes (DNA and RNA polymerases) and by solid-phase chemical synthesis. The chemical methods also enable the generation of altered nucleic acids that are not found in nature, for example peptide nucleic acids.

Molecular Composition and Size

Nucleic acids can vary in size, but are generally very large molecules. Indeed, DNA molecules are probably the largest individual molecules known. Well-studied biological nucleic acid molecules range in size from 21 nucleotides (small interfering RNA) to large chromosomes (human chromosome 1 is a single molecule that contains 247 million base pairs). In most cases, naturally occurring DNA molecules are double-stranded and RNA molecules are single-stranded. There are numerous exceptions, however—some viruses have genomes made of double-stranded RNA and other viruses have single-stranded DNA genomes, and, in some circumstances, nucleic acid structures with three or four strands can form.

Nucleic acids are linear polymers (chains) of nucleotides. Each nucleotide consists of three components: a purine or pyrimidine nucleobase (sometimes termed *nitrogenous base* or simply *base*), a pentose sugar; and a phosphate group. The substructure consisting of a nucleobase plus sugar is termed a nucleoside. Nucleic acid types differ in thc structure of the sugar in their nucleotides - DNA contains 2'-deoxyribose while RNA contains ribose (where the only difference is the presence of a hydroxyl group). Also, the nucleobases found in the two nucleic acid types are different: adenine, cytosine, and guanine are found in both RNA and DNA, while thymine occurs in DNA and uracil occurs in RNA. The sugars and phosphates in nucleic acids are connected to each other in an alternating chain (sugar-phosphate backbone) through phosphodiester linkages. In conventional nomenclature, the carbons to which the phosphate groups attach are the 3'-end and the 5'-end carbons of the sugar. This gives nucleic acids directionality, and the ends of nucleic acid molecules are referred to as 5'-end and 3'-end. The nucleobases are joined to the sugars via an N-glycosidic linkage involving a nucleobase ring nitrogen (N-1 for pyrimidines and N-9 for purines) and the 1' carbon of the pentose sugar ring. Non-standard nucleosides are also found in both RNA and DNA and usually arise from modification of the standard nucleosides within the DNA molecule or the primary (initial) RNA transcript. Transfer RNA (tRNA) molecules contain a particularly large number of modified nucleosides.

Topology

Double-stranded nucleic acids are made up of complementary sequences, in which extensive Watson-Crick base pairing results in the formation of a highly repeated and quite uniform double-helical three-dimensional structure. In contrast, single-stranded RNA and DNA molecules are not constrained to a regular double helix, and can adopt highly complex three-dimensional structures that are based on short stretches of intramolecular base-paired sequences that include both Watson-Crick and noncanonical base pairs, as well as a wide range of complex tertiary interactions. Nucleic acid molecules are usually unbranched, and may occur as linear and circular molecules. For example, bacterial chromosomes, plasmids, mitochondrial DNA and chloroplast DNA are usually circular double-stranded DNA molecules, while chromosomes of the eukaryotic nucleus are usually linear double-stranded DNA molecules. Most RNA molecules are linear, single-stranded molecules, but both circular and branched molecules can result from RNA splicing reactions.

Nucleic Acid Sequences

One DNA or RNA molecule differs from another primarily in the sequence of nucleotides. Nucleotide sequences are of great importance in biology, since they carry the ultimate instructions that encode all biological molecules, molecular assemblies, subcellular and cellular structures, organs and organisms, and directly enable cognition, memory and behaviour. Enormous efforts have gone into the development of experimental methods to determine the nucleotide sequence of biological DNA and RNA molecules, and today hundreds of millions of nucleotides are sequenced daily at genome centres and smaller laboratories worldwide.

Types of Nucleic Acids

Deoxyribonucleic Acid: Deoxyribonucleic acid is a nucleic acid that contains the genetic instructions used in the development and functioning of all known living organisms. The main role of DNA molecules is the long-term storage of information and DNA is often compared to a set of blueprints, since it contains the instructions needed to construct other components of cells, such as proteins and RNA molecules. The DNA segments that carry this genetic information are called genes, but other DNA sequences have structural purposes, or are involved in regulating the use of this genetic information.

Ribonucleic Acid: Ribonucleic acid (RNA) functions in converting genetic information from genes into the amino acid sequences of proteins. The three universal types of RNA include transfer RNA (tRNA), messenger RNA (mRNA), and ribosomal RNA (rRNA). Messenger RNA acts to carry genetic sequence information between DNA and ribosomes, directing protein synthesis. Ribosomal RNA is a major component of the ribosome, and catalyzes peptide bond formation. Transfer RNA serves as the carrier molecule for amino acids to be used in protein synthesis, and is responsible for decoding the mRNA. In addition, many other classes of RNA are now known.

Artificial Nucleic Acid Analogs: Artificial nucleic acid analogs have been designed and synthesized by chemists, and include peptide nucleic acid, morpholino- and locked nucleic acid, as well as glycol nucleic acid and threose nucleic acid. Each of these is distinguished from naturally-occurring DNA or RNA by changes to the backbone of the molecule.

DNA

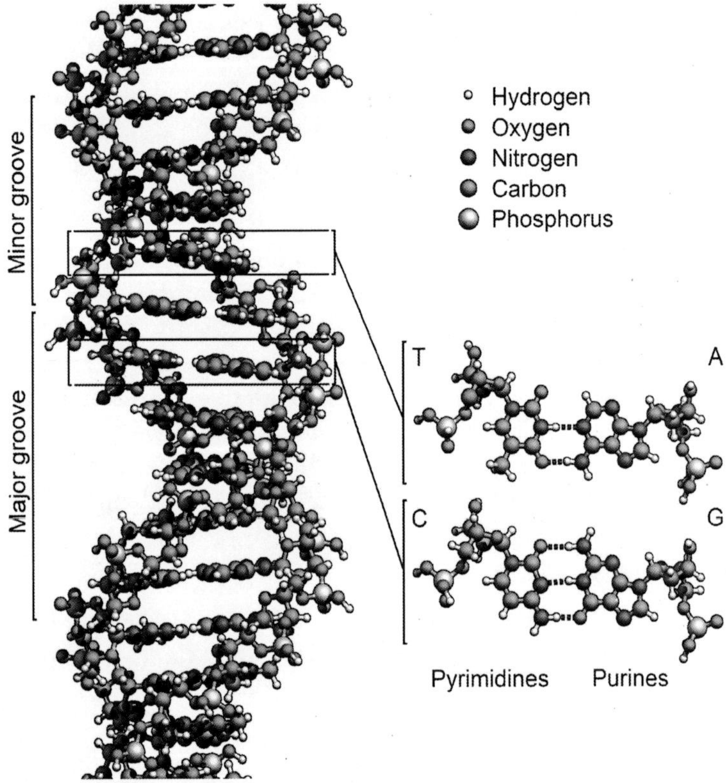

Figure: *The structure of the DNA double helix. The atoms in the structure are colour coded by element and the detailed structure of two base pairs is shown in the bottom right.*

Deoxyribonucleic acid or DNA, is a nucleic acid that contains the genetic instructions used in the development and functioning of all known living organisms (with the exception of RNA viruses). The main role of DNA molecules is the long-term storage of information. DNA is often compared to a set of blueprints, like a recipe or a code, since it contains the instructions needed to construct other components of cells, such as proteins and RNA molecules. The DNA segments that carry this genetic information are called genes, but other DNA sequences have structural purposes, or are involved in regulating the use of this genetic information.

DNA consists of two long polymers of simple units called nucleotides, with backbones made of sugars and phosphate groups joined by ester bonds. These two strands run in opposite directions to each other and are therefore antiparallel. Attached to each sugar is one of four types of molecules called bases. It is the sequence of these four bases along the backbone that encodes information. This information is read using the genetic code, which specifies the sequence of the amino acids within proteins. The code is read by copying stretches of DNA into the related nucleic acid RNA, in a process called transcription.

Within cells, DNA is organized into long structures called chromosomes. These chromosomes are duplicated before cells divide, in a process called DNA replication. Eukaryotic organisms (animals, plants, fungi, and protists) store most of their DNA inside the cell nucleus and some of their DNA in organelles, such as mitochondria or chloroplasts. In contrast, prokaryotes (bacteria and archaea) store their DNA nly in the cytplasm Within the chromosoes, chromatin prteins such as histones compact and organize DNA. These compact structures guide the interactions between DNA and other proteins, helping control which parts of the DNA are transcribed.

Properties

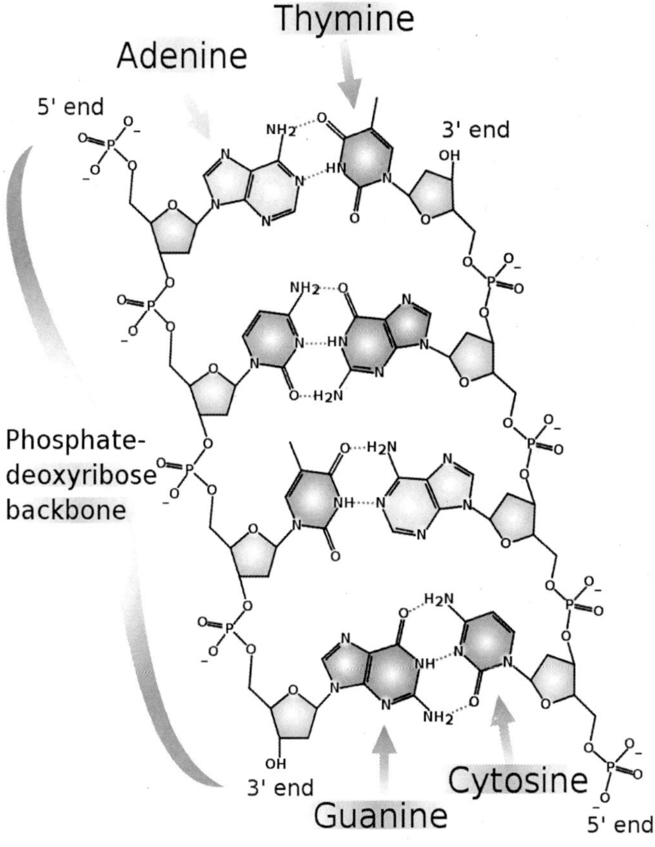

Figure: *Chemical structure of DNA. Hydrogen bonds shown as dotted lines.*

DNA is a long polymer made from repeating units called nucleotides. As first discovered by James D. Watson and Francis Crick, the structure of DNA of all species comprises two helical chains each coiled round the same axis, and each with a pitch of 34 Ångströms (3.4 nanometres) and a radius of 10 Ångströms (1.0 nanometres). According to another study, when measured in a particular solution, the DNA chain measured 22 to 26 Ångströms wide (2.2 to 2.6 nanometres), and one nucleotide unit measured 3.3 Å (0.33 nm) long. Although each individual repeating unit is very small, DNA polymers can be very large molecules containing millions of nucleotides. For instance, the largest human chromosome, chromosome number 1, is approximately 220 million base pairs long.

In living organisms, DNA does not usually exist as a single molecule, but instead as a pair of molecules that are held tightly together. These two long strands entwine like vines, in the shape of a double helix. The nucleotide repeats contain both the segment of the backbone of the molecule, which holds the chain together, and a base, which interacts with the other DNA strand in the helix. A base linked to a sugar is called a nucleoside and a base linked to a sugar and one or more phosphae groups is called a nucleotide. If multiple nucleotides are linked together, as in DNA, this polymer is called a polynucleotide. The backbone of the DNA strand is made from alternating phosphate and sugar residues.

The sugar in DNA is 2-deoxyribose, which is a pentose (five-carbon) sugar. The sugars are joined together by phosphate groups that form phosphodiester bonds between the third and fifth carbon atoms of adjacent sugar rings. These asymmetric bonds mean a strand of DNA has a direction. In a double helix the direction of the nucleotides in one strand is opposite to their direction in the other strand: the strands are *antiparallel*. The asymmetric ends of DNA strands are called the 52 (*five prime*) and 32 (*three prime*) ends, with the 5' end having a terminal phosphate group and the 3' end a terminal hydroxyl group. One major difference between DNA and RNA is the sugar, with the 2-deoxyribose in DNA being replaced by the alternative pentose sugar ribose in RNA.

The DNA double helix is stabilized primarily by two forces: hydrogen bonds between nucleotides and base-stacking interactions among the aromatic bases. In the aqueous environment of the cell, the conjugated ð bonds of nucleotide bases align perpendicular to the axis of the DNA molecule, minimizing their interaction with the solvation shell and therefore, the Gibbs free energy. The four bases found in DNA are adenine (abbreviated A), cytosine (C), guanine (G) and thymine (T). These four bases are attached to the sugar/ phosphate to form the complete nucleotide, as shown for adenosine monophosphate.

These bases are classified into two types; adenine and guanine are fused five- and six-membered heterocyclic compounds called purines, while cytosine and thymine are six-membered rings called pyrimidines. A fifth pyrimidine base, called uracil (U), usually takes the place of thymine in RNA and differs from thymine by lacking a methyl group on its ring. Uracil is not usually found in DNA, occurring only as a breakdown product of cytosine. In addition to RNA and DNA, a large number of artificial nucleic acid analogues have also been created to study the proprieties of nucleic acids, or for use in biotechnology.

Grooves: Twin helical strands form the DNA backbone. Another double helix may be found by tracing the spaces, or grooves, between the strands. These voids are adjacent to the base pairs and may provide a binding site. As the strands are not directly opposite each other, the grooves are unequally sized. One groove, the major groove, is 22 Å wide and the other, the minor groove, is 12 Å wide. The narrowness of the minor groove means that the edges of the bases are more accessible in the major groove. As a result, proteins like transcription factors that can bind to specific sequences in double-stranded DNA usually make contacts to the sides of the bases exposed in the major groove. This situation varies in unusual conformations of DNA within the cell, but the major and minor grooves are always named to reflect the differences in size that would be seen if the DNA is twisted back into the ordinary B form.

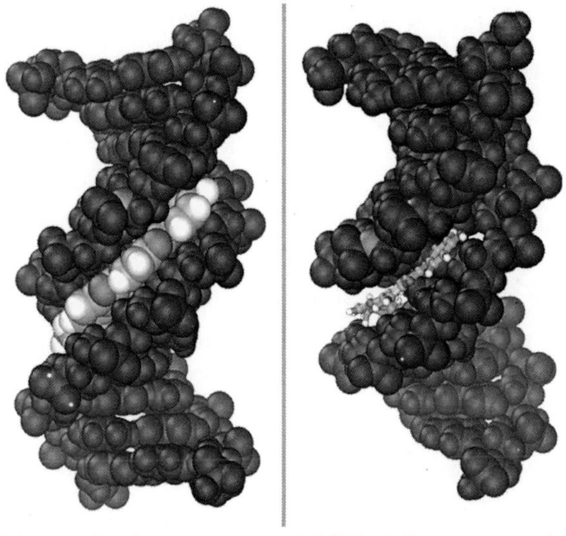

Figure: *Major and minor grooves of DNA. Minor groove is a binding site for the dye Hoechst 33258.*

Base Pairing: Each type of base on one strand forms a bond with just one type of base on the other strand. This is called complementary base pairing. Here, purines form hydrogen bonds to pyrimidines, with A bonding only to T, and C bonding only to G. This arrangement of two nucleotides binding together across the double helix is called a base pair. As hydrogen bonds are not covalent, they can be broken and rejoined relativelyeasily. The two strands of DNA in a double helix can therefore be pulled apart like a zipper, either by a mechanical force or high temperature. As a result of this complementarity, all the information in the double-stranded sequence of a DNA helix is duplicated on each strand, which is vital in DNA replication. Indeed, this reversible and specific interaction between complementary base pairs is critical for all the functions of DNA in living organisms.

The two types of base pairs form different numbers of hydrogen bonds, AT forming two hydrogen bonds, and GC forming three hydrogen bonds. DNA with high GC-content is more stable than DNA with low GC-content, due t the added stability of an additional hydrogen bond. As a result, it is both the percentage of GC base pairs and the overall length of a DNA double helix that determine the strength of the association between the two strands of DNA.

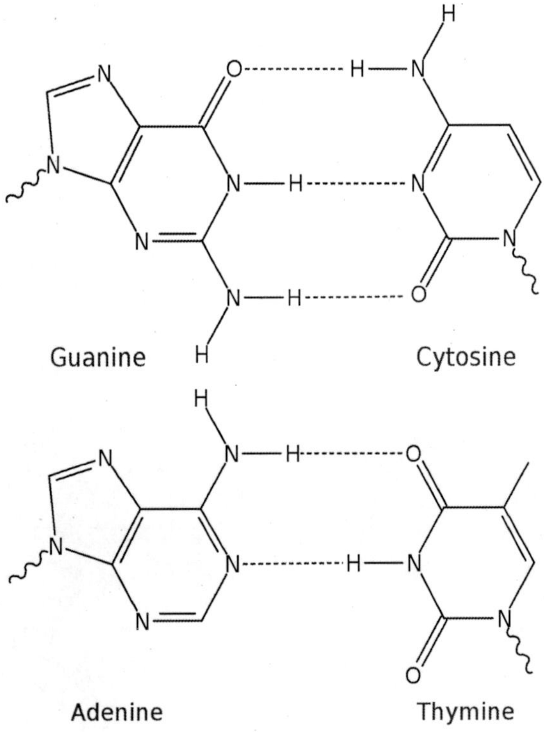

Figure: *Top, a GC base pair with three hydrogen bonds. Bottom, an AT base pair with two hydrogen bonds. Non-covalent hydrogen bonds between the pairs are shown as dashed lines.*

Long DNA helices with a high GC content have stronger-interacting strands, while short helices with high AT content have weaker-interacting strands. In biology, parts of the DNA double helix that need to separate easily, such as the TATAAT Pribnow box in some promoters, tend to have a high AT content, making the strands easier to pull apart. In the laboratory, thestrength of this interaction can be measured by finding the temperature required to break the hydrogen bonds, their melting temperature (also called *Tm* value). When all the base pairs in a DNA double helix melt, the strands separate and exist in solution as two entirely independent molecules. These single-stranded DNA molecules (*ssDNA*) have no single common shape, but some conformations are more stable than others.

Sense and Antisense: A DNA sequence is called "sense" if its sequence is the same as that of a messenger RNA copy that is translated into protein. The sequence on the opposite strand is called the "antisense" sequence. Both sense and antisense sequences can exist on different parts of the same strand of DNA (i.e. both strands contain both sense and antisense sequences). In both prokaryotes and eukaryotes, antisense RNA sequences are produced, but the functions of these RNAs are not entirely clear. One proposal is that antisense RNAs are involved in regulating gene exression through RNA-RNA base pairing.

A few DNA sequences in prokaryotes and eukaryotes, and more in plasmids and viruses, blur the distinction between sense and antisense strands by having overlapping genes. In these cases, some DNA sequences do double duty, encoding one protein when read along one strand, and a second protein when read in the opposite direction along the other strand. In bacteria, this overlap may be involved in the regulation of gene transcription, while in viruses, overlapping genes increase the amount of information that can be encoded within the small viral genome.

Supercoiling: DNA can be twisted like a rope in a process called DNA supercoiling. With DNA in its "relaxed" state, a strand usually circles the axis of the double helix once every 10.4 base pairs, but if the DNA is twisted the strands become more tightly or more loosely wound. If the DNA is twisted in the direction of the helix, this is positive supercoiling, and the bases are held more tightly together. If they are twisted in the opposite direction, this is negative supercoiling, and the bases come apart more easily. In nature, most DNA has slight negative supercoiling that is introduced by enzymes called topoisomerases. These enzymes are also needed to relieve the twisting stresses introduced into DNA strands during processes such as transcription and DNA replication.

Alternate DNA Structures: DNA exists in many possible conformations that include A-DNA, B-DNA, and Z-DNA forms, although, only B-DNA and Z-DNA have been directly observed in functional organisms. The conformation that DNA adopts depends on the hydration level, DNA sequence, the amount and direction of supercoiling, chemical modifications of the bases, the type and concentration of metal ions, as well as the presence of polyamines in solution.

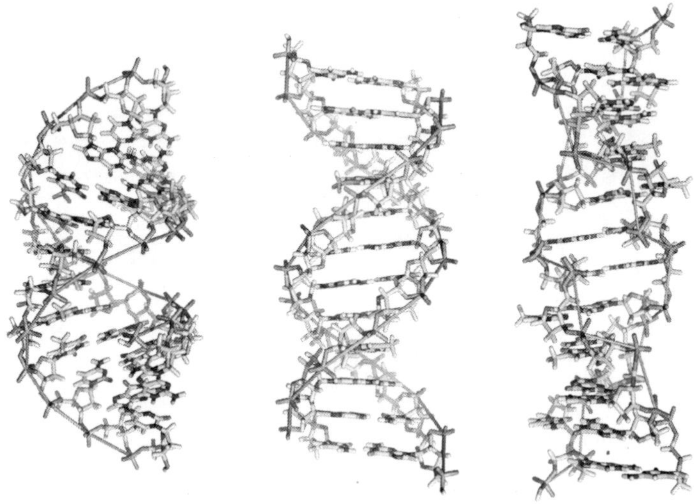

Figure: *From left to right, the structures of A, B and Z DNA*

The first published reports of A-DNA X-ray diffraction patterns— and also B-DNA used analyses based on Patterson transforms that provided only a limited amount of structural information for oriented fibres of DNA. An alternate analysis was then proposed by Wilkins *et al.*, in 1953, for the *in vivo* B-DNA X-ray diffraction/scattering patterns of highly hydrated DNA fibres in terms of squares of Bessel functions. In the same journal, James D. Watson and Francis Crick presented their molecular modelling analysis of the DNA X-ray diffraction patterns to suggest that the structure was a double-helix.

Although the 'B-DNA form' is most common under the conditions found in cells, it is not a well-defined conformation but a family of related DNA conformations that occur at the high hydration levels present in living cells. Their corresponding X-ray diffraction and scattering patterns are characteristic of molecular paracrystals with a significant degree of disorder. Compared to B-DNA, the A-DNA form is a wider right-handed spiral, with a shallow, wide minor groove and a narrower, deeper major groove. The A form occurs under non-physiological conditions in partially dehydrated samples of DNA, while in the cell it may be produced in hybrid pairings of DNA and RNA strands, as well as in enzyme-DNA complexes. Segments of DNA where the bases have been chemically modified by methylation may undergo a larger change in conformation and adopt the Z form. Here, the strands turn about the helical axis in a left-handed spiral, the opposite of the more common B form. These unusual structures can be recognized by specific Z-DNA binding proteins and may be involved in the regulation of transcription.

Alternate DNA Chemistry: For a number of years exobiologists have proposed the existence of a shadow biosphere, a postulated microbial biosphere of Earth that uses radically different biochemical and molecular processes than currently known life. One of the proposals was the existence of lifeforms that use arsenic instead of phosphorus in DNA.

A December 2010 NASA press conference revealed that the bacterium GFAJ-1, which has evolved in an arsenic-rich environment, is the first terrestrial lifeform found which may have this ability. The bacterium was found in Mono Lake, east of Yosemite National Park. GFAJ-1 is a rod-shaped extremophile bacterium in the family Halomonadaceae that, when starved of phosphorus, may be capable of incorporating the usually poisonous element arsenic in its DNA. This discovery may lend weight to the long-standing idea that extraterrestrial life could have a different chemical makeup from life on Earth. The research was carried out by a team led by Felisa Wolfe-Simon, a geomicrobiologist and geobiochemist, a Postdoctoral Fellow of the NASA Astrobiology Institute with Arizona State University. This finding has, however, faced strong criticism from the scientific community; scientists have argued that there is no evidence that arsenic is actually incorporated into biomolecules. Independent conformation of this finding has also not yet been possible.

Quadruplex Structures: At the ends of the linear chromosomes are specialized regions of DNA called telomeres.

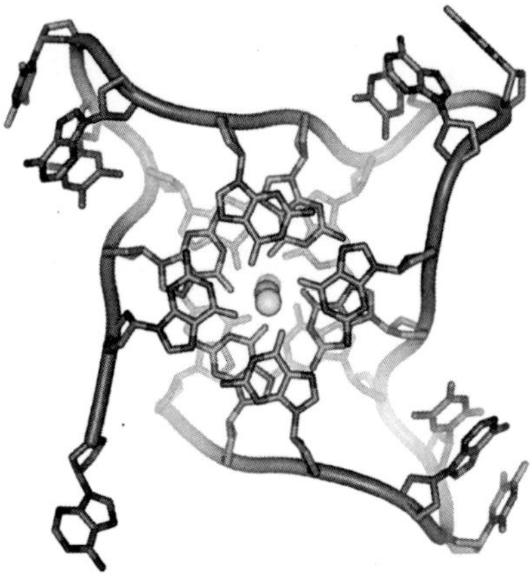

Figure: *DNA quadruplex formed by telomere repeats. The looped conformation of the DNA backbone is very different from the typical DNA helix.*

The main function of these regions is to allow the cell to replicate chromosome ends using the enzyme telomerase, as the enzymes that normally replicate DNA cannot copy the extreme 32 ends of chromosomes. These specialized chromosome caps also help protect the DNA ends, and stop the DNA repair systems in the cell from treating them as damage to be corrected. In human cells, telomeres are usually lengths of single-stranded DNA containing several thousand repeats of a simple TTAGGG sequence. These guanine-rich sequences may stabilize chromosome ends by forming structures of stacked sets of four-base units, rather than the usual base pairs found in other DNA molecules. Here, four guanine bases form a flat plate and these flat four-base units then stack on top of each other, to form a stable *G-quadruplex* structure. These structures are stabilized by hydrogen bonding between the edges of the bases and chelation of a metal ion in the centre of each four-base unit. Other structures can also be formed, with the central set of four bases coming from either a single strand folded around the bases, or several different parallel strands, each contributing one base to the central structure. In addition to these stacked structures, telomeres also form large loop structures called telomere loops, or T-loops. Here, the single-stranded DNA curls around in a long circle stabilized by telomere-binding proteins. At the very end of the T-loop, the single-stranded telomere DNA is held onto a region of double-stranded DNA by the telomere strand disrupting the double-helical DNA and base pairing to one of the two strands. This triple-stranded structure is called a displacement loop or D-loop.

Branched DNA: In DNA fraying occurs when non-complementary regions exist at the end of an otherwise complementary double-strand of DNA. However, branched DNA can occur if a third strand of DNA is introduced and contains adjoining regions able to hybridize with the frayed regions of the pre-existing double-strand. Although the simplest example of branched DNA involves only three strands of DNA, complexes involving additional strands and multiple branches are also possible. Branched DNA can be used in nanotechnology to construct geometric shapes, see the section on uses in technology below.

Vibration: DNA may carry out low-frequency collective motion as observed by the Raman spectroscopy and analyzed with a quasi-continuum model.

Chemical Modifications

Base Modifications: The expression of genes is influenced by how the DNA is packaged in chromosomes, in a structure called chromatin. Base modifications can be involved in packaging, with regions that have low or no gene expression usually containing high levels of methylation of cytosine bases. For example, cytosine methylation, produces 5-methylcytosine, which is important for X-chromosome inactivation. The average level of methylation varies between organisms - the worm *Caenorhabditis elegans* lacks cytosine methylation, while vertebrates have higher levels, with up to 1% of their DNA containing 5-methylcytosine. Despite the importance of 5-methylcytosine, it can deaminate to leave a thymine base, so methylated cytosines are particularly prone to mutations. Other base modifications include adenine methylation in bacteria, the presence of 5-hydroxymethylcytosine in the brain, and the glycosylation of uracil to produce the "J-base" in kinetoplastids.

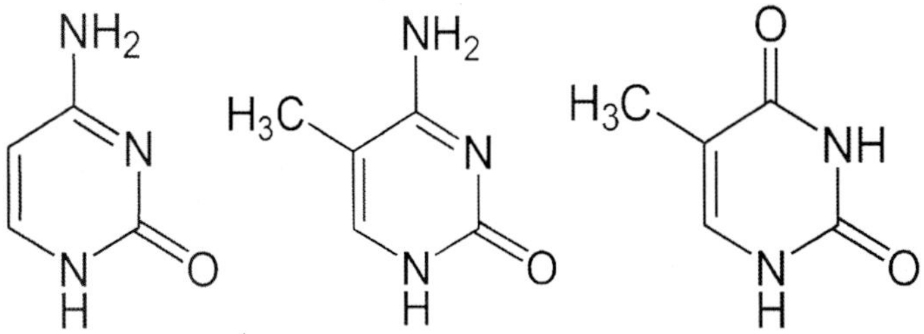

Figure: *Structure of cytosine with and without the 5-methyl group. Deamination converts 5-methylcytosine into thymine.*

Damage: DNA can be damaged by many sorts of mutagens, which change the DNA sequence. Mutagens include oxidizing agents, alkylating agents and also high-energy electromagnetic radiation such as ultraviolet light and X-rays. The type of DNA damage produced depends on the type of mutagen. For example, UV light can damage DNA by producing thymine dimers, which are cross-links between pyrimidine bases. On the other hand, oxidants such as free radicals or hydrogen peroxide produce multiple forms of damage, including base modifications, particularly of guanosine, and double-strand breaks. A typical human cell contains about 150,000 bases that have suffered oxidative damage. Of these oxidative lesions, the most dangerous are double-strand breaks, as these are difficult to repair and can produce point mutations, insertions and deletions from the DNA sequence, as well as chromosomal translocations.

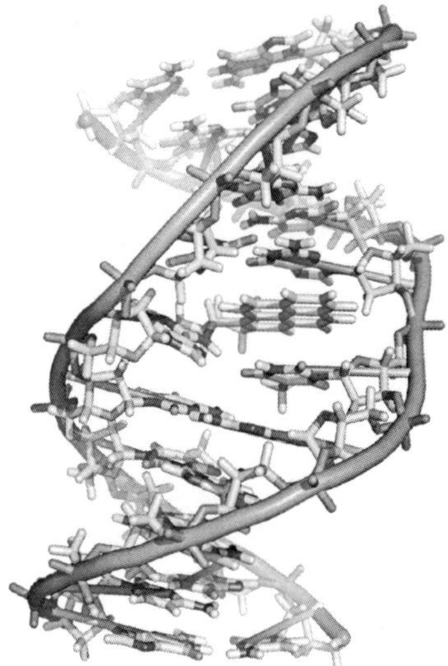

Figure: *A covalent adduct between a metabolically activated form of benzo[a]pyrene, the major mutagen in tobacco smoke, and DNA*

Many mutagens fit into the space between two adjacent base pairs, this is called *intercalation*. Most intercalators are aromatic and planar molecules;

examples include ethidium bromide, daunomycin, and doxorubicin. In order for an intercalator to fit between base pairs, the bases must separate, distorting the DNA strands by unwinding of the double helix. This inhibits both transcription and DNA replication, causing toxicity and mutations. As a result, DNA intercalators are often carcinogens, and benzo[*a*]pyrene diol epoxide, acridines, aflatoxin and ethidium bromide are well-known examples. Nevertheless, due to their ability to inhibit DNA transcription and replication, other similar toxins are also used in chemotherapy to inhibit rapidly growing cancer cells.

Biological Functions: DNA usually occurs as linear chromosomes in eukaryotes, and circular chromosomes in prokaryotes. The set of chromosomes in a cell makes up its genome; the human genome has approximately 3 billion base pairs of DNA arranged into 46 chromosomes. The information carried by DNA is held in the sequence of pieces of DNA called genes. Transmission of genetic information in genes is achieved via complementary base pairing. For example, in transcription, when a cell uses the information in a gene, the DNA sequence is copied into a complementary RNA sequence through the attraction between the DNA and the correct RNA nucleotides. Usually, this RNA copy is then used to make a matching protein sequence in a process called translation, which depends on the same interaction between RNA nucleotides. In alternative fashion, a cell may simply copy its genetic information in a process called DNA replication. The details of these functions are covered in other articles; here we focus on the interactions between DNA and other molecules that mediate the function of the genome.

Genes and Genomes: Genomic DNA is tightly and orderly packed in the process called DNA condensation to fit the small available volumes of the cell. In eukaryotes, DNA is located in the cell nucleus, as well as small amounts in mitochondria and chloroplasts. In prokaryotes, the DNA is held within an irregularly shaped body in the cytoplasm called the nucleoid. The genetic information in a genome is held within genes, and the complete set of this information in an organism is called its genotype. A gene is a unit of heredity and is a region of DNA that influences a particular characteristic in an organism. Genes contain an open reading frame that can be transcribed, as well as regulatory sequences such as promoters and enhancers, which control the transcription of the open reading frame.

In many species, only a small fraction of the total sequence of the genome encodes protein. For example, only about 1.5% of the human genome consists of protein-coding exons, with over 50% of human DNA consisting of non-coding repetitive sequences. The reasons for the presence of so much noncoding DNA in eukaryotic genomes and the extraordinary differences in genome size, or *C-value*, among species represent a long-standing puzzle known as the "C-value enigma". However, DNA sequences that do not code protein may still encode functional non-coding RNA molecules, which are involved in the regulation of gene expression. Some noncoding DNA sequences play structural roles in chromosomes. Telomeres and centromeres typically contain few genes, but are important for the function and stability of chromosomes. An abundant form of noncoding DNA in humans are pseudogenes, which are copies of genes that have been disabled by mutation. These sequences are usually just molecular fossils, although they can occasionally serve as raw genetic material for the creation of new genes through the process of gene duplication and divergence.

Transcription and Translation: A gene is a sequence of DNA that contains genetic information and can influence the phenotype of an organism. Within a gene, the sequence of bases along a DNA strand defines a messenger RNA sequence, which then defines one or more protein sequences. The relationship between the nucleotide sequences of genes and the amino-acid sequences of proteins is determined by the rules of translation, known collectively as the genetic code. The genetic code consists of three-letter 'words' called *codons* formed from a sequence of three nucleotides (e.g. ACT, CAG, TTT). In transcription, the codons of a gene are copied into messenger RNA by RNA polymerase. This RNA copy is then decoded by a ribosome that reads the RNA sequence by base-pairing the messenger RNA to transfer RNA, which carries amino acids. Since there are 4 bases in 3-letter combinations, there are 64 possible codons (43 combinations). These encode the twenty standard amino acids, giving most amino acids more than one possible codon. There are also three 'stop' or 'nonsense' codons signifying the end of the coding region; these are the TAA, TGA and TAG codons.

Replication: Cell division is essential for an organism to grow, but, when a cell divides, it must replicate the DNA in its genome so that the two daughter cells have the same genetic information as their parent.

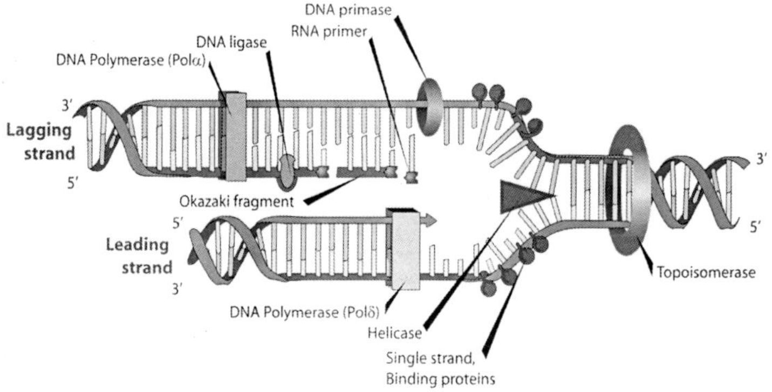

Figure: *DNA replication. The double helix is unwound by a helicase and topoisomerase. Next, one DNA polymerase produces the leading strand copy. Another DNA polymerase binds to the lagging strand. This enzyme makes discontinuous segments (called Okazaki fragments) before DNA ligase joins them together.*

The double-stranded structure of DNA provides a simple mechanism for DNA replication. Here, the two strands are separated and then each strand's complementary DNA sequence is recreated by an enzyme called DNA polymerase. This enzyme makes the complementary strand by finding the correct base through complementary base pairing, and bonding it onto the original strand. As DNA polymerases can only extend a DNA strand in a 52 to 32 direction, different mechanisms are used to copy the antiparallel strands of the double helix. In this way, the base on the old strand dictates which base appears on the new strand, and the cell ends up with a perfect copy of its DNA.

Interactions with Proteins: All the functions of DNA depend on interactions with proteins. These protein interactions can be non-specific, or the protein can bind specifically to a single DNA sequence. Enzymes can also bind to DNA and of these, the polymerases that copy the DNA base sequence in transcription and DNA replication are particularly important.

DNA-Binding Proteins: Structural proteins that bind DNA are well-understood examples of non-specific DNA-protein interactions. Within chromosomes, DNA is held in complexes with structural proteins. These proteins organize the DNA into a compact structure called chromatin. In eukaryotes this structure involves DNA binding to a complex of small basic proteins called histones, while in prokaryotes multiple types of proteins are involved. The histones form a disk-shaped complex called a nucleosome, which contains two complete turns of double-stranded DNA wrapped around its surface. These non-specific interactions are formed through basic residues in the histones making ionic bonds to the acidic sugar-phosphate backbone of the DNA, and are therefore largely independent of the base sequence.

Chemical modifications of these basic amino acid residues include methylation, phosphorylation and acetylation. These chemical changes alter the strength of the interaction between the DNA and the histones, making the DNA more or less accessible to transcription factors and changing the rate of transcription. Other non-specific DNA-binding proteins in chromatin include the high-mobility group proteins, which bind to bent or distorted DNA. These proteins are important in bending arrays of nucleosomes and arranging them into the larger structures that make up chromosomes. A distinct group of DNA-binding proteins are the DNA-binding proteins that specifically bind single-stranded DNA.

In humans, replication protein A is the best-understood member of this family and is used in processes where the double helix is separated, including DNA replication, recombination and DNA repair. These binding proteins seem to stabilize single-stranded DNA and protect it from forming stem-loops or being degraded by nucleases. In contrast, other proteins have evolved to bind to particular DNA sequences. The most intensively studied of these are the various transcription factors, which are proteins that regulate transcription. Each transcription factor binds to one particular set of DNA sequences and activates or inhibits the transcription of genes that have these sequences close to their promoters.

The transcription factors do this in two ways. Firstly, they can bind the RNA polymerase responsible for transcription, either directly or through other mediator proteins; this locates the polymerase at the promoter and allows it to begin transcription. Alternatively, transcription factors can bind enzymes that modify the histones at the promoter; this will change the accessibility of the DNA template to the polymerase. As these DNA targets can occur throughout an organism's genome, changes in the activity of one type of transcription factor can affect thousands of genes.

Consequently, these proteins are often the targets of the signal transduction processes that control responses to environmental changes or cellular differentiation and development. The specificity of these transcription factors' interactions with DNA come from the proteins making multiple contacts to the edges of the DNA bases, allowing them to "read" the DNA sequence. Most of these base-interactions are made in the major groove, where the bases are most accessible.

DNA-Modifying Enzymes

Nucleases and Ligases: Nucleases are enzymes that cut DNA strands by catalyzing the hydrolysis of the phosphodiester bonds. Nucleases that hydrolyse nucleotides from the ends of DNA strands are called exonucleases, while endonucleases cut within strands. The most frequently used nucleases in molecular biology are the restriction endonucleases, which cut DNA at specific sequences. For instance, the EcoRV enzyme shown to the left recognizes the 6-base sequence 52 -GAT | ATC-32 and makes a cut at the vertical line. In nature, these enzymes protect bacteria against phage infection by digesting the phage DNA when it enters the bacterial cell, acting as part of the restriction modification system. In technology, these sequence-specific nucleases are used in molecular cloning and DNA fingerprinting.

Enzymes called DNA ligases can rejoin cut or broken DNA strands. Ligases are particularly important in lagging strand DNA replication, as they join together the short segments of DNA produced at the replication fork into a complete copy of the DNA template. They are also used in DNA repair and genetic recombination.

Topoisomerases and Helicases: Topoisomerases are enzymes with both nuclease and ligase activity. These proteins change the amount of supercoiling in DNA. Some of these enzymes work by cutting the DNA helix and allowing one section to rotate, thereby reducing its level of supercoiling; the enzyme then seals the DNA break. Other types of these enzymes are capable of cutting one DNA helix and then passing a second strand of DNA through this break, before rejoining the helix. Topoisomerases are required for many processes involving DNA, such as DNA replication and transcription. Helicases are proteins that are a type of molecular motor. They use the chemical energy in nucleoside triphosphates, predominantly ATP, to break hydrogen bonds between bases and unwind the DNA double helix into single strands. These enzymes are essential for most processes where enzymes need to access the DNA bases.

Polymerases: Polymerases are enzymes that synthesize polynucleotide chains from nucleoside triphosphates. The sequence of their products are copies of existing polynucleotide chains - which are called *templates*. These enzymes function by adding nucleotides onto the 32 hydroxyl group of the previous nucleotide in a DNA strand. As a consequence, all polymerases work in a 52 to 32 direction. In the active site of these enzymes, the incoming nucleoside triphosphate base-pairs to the template: this allows polymerases to accurately synthesize the complementary strand of their template. Polymerases are classified according to the type of template that they use. In DNA replication, a DNA-dependent DNA polymerase makes a copy of a DNA sequence. Accuracy is vital in this process, so many of these polymerases have a proofreading activity. Here, the polymerase recognizes the occasional mistakes in the synthesis reaction by the lack of base pairing between the mismatched nucleotides. If a mismatch is detected, a 32 to 52 exonuclease activity is activated and the incorrect base removed. In most organisms, DNA polymerases function in a large complex called the replisome that contains multiple accessory subunits, such as the DNA clamp or helicases.

RNA-dependent DNA polymerases are a specialized class of polymerases that copy the sequence of an RNA strand into DNA. They include reverse transcriptase, which is a viral enzyme involved in the infection of cells by retroviruses, and telomerase, which is required for the replication of telomeres. Telomerase is an unusual polymerase because it contains its own RNA template as part of its structure. Transcription is carried out by a DNA-dependent RNA polymerase that copies the sequence of a DNA strand into RNA. To begin transcribing a gene, the RNA polymerase binds to a sequence of DNA called a promoter and separates the DNA strands. It then copies the gene sequence into a messenger RNA transcript until it reaches a region of DNA called the terminator, where it halts and detaches from the DNA. As with human DNA-dependent DNA polymerases, RNA polymerase II, the enzyme that transcribes most of the genes in the human genome, operates as part of a large protein complex with multiple regulatory and accessory subunits.

Genetic Recombination

DNA helix usually does not interact with other segments of DNA, and in human cells the different chromosomes even occupy separate areas in the nucleus called "chromosome territories". This physical separation of different chromosomes is important for the ability of DNA to function as a stable repository for information, as one of the few times chromosomes interact is during chromosomal crossover when they recombine. Chromosomal crossover is when two DNA helices break, swap a section and then rejoin.

Recombination allows chromosomes to exchange genetic information and produces new combinations of genes, which increases the efficiency of natural selection and can be important in the rapid evolution of new proteins. Genetic recombination can also be involved in DNA repair, particularly in the cell's response to double-strand breaks. The most common form of chromosomal crossover is homologous recombination, where the two chromosomes involved share very similar sequences. Non-homologous recombination can be damaging to cells, as it can produce chromosomal translocations and genetic abnormalities.

The recombination reaction is catalyzed by enzymes known as recombinases, such as RAD51. The first step in recombination is a double-stranded break either caused by an endonuclease or damage to the DNA. A series of steps catalyzed in part by the recombinase then leads to joining of the two helices by at least one Holliday junction, in which a segment of a single strand in each helix is annealed to the complementary strand in the other helix. The Holliday junction is a tetrahedral junction structure that can be moved along the pair of chromosomes, swapping one strand for another. The recombination reaction is then halted by cleavage of the junction and re-ligation of the released DNA.

Evolution

DNA contains the genetic information that allows all modern living things to function, grow and reproduce. However, it is unclear how long in the 4-billion-year history of life DNA has performed this function, as it has been proposed that the earliest forms of life may have used RNA as their genetic material. RNA may have acted as the central part of early cell metabolism as it can both transmit genetic information and carry out catalysis as part of ribozymes. This ancient RNA world where nucleic acid would have been used for both catalysis and genetics may have influenced the evolution of the current genetic code based on four nucleotide bases.

This would occur, since the number of different bases in such an organism is a trade-off between a small number of bases increasing replication accuracy and a large number of bases increasing the catalytic efficiency of ribozymes.

However, there is no direct evidence of ancient genetic systems, as recovery of DNA from most fossils is impossible. This is because DNA will survive in the environment for less than one million years and slowly degrades into short fragments in solution. Claims for older DNA have been made, most notably a report of the isolation of a viable bacterium from a salt crystal 250 million years old, but these claims are controversial.

Uses in Technology

Genetic Engineering: Methods have been developed to purify DNA from organisms, such as phenol-chloroform extraction, and to manipulate it in the laboratory, such as restriction digests and the polymerase chain reaction. Modern biology and biochemistry make intensive use of these techniques in recombinant DNA technology. Recombinant DNA is a man-made DNA sequence that has been assembled from other DNA sequences. They can be transformed into organisms in the form of plasmids or in the appropriate format, by using a viral vector. The genetically modified organisms produced can be used to produce products such as recombinant proteins, used in medical research, or be grown in agriculture.

Forensics: Forensic scientists can use DNA in blood, semen, skin, saliva or hair found at a crime scene to identify a matching DNA of an individual, such as a perpetrator. This process is formally termed DNA profiling, but may also be called "genetic fingerprinting". In DNA profiling, the lengths of variable sections of repetitive DNA, such as short tandem repeats and minisatellites, are compared between people. This method is usually an extremely reliable technique for identifying a matching DNA. However, identification can be complicated if the scene is contaminated with DNA from several people. DNA profiling was developed in 1984 by British geneticist Sir Alec Jeffreys, and first used in forensic science to convict Colin Pitchfork in the 1988 Enderby murders case. People convicted of certain types of crimes may be required to provide a sample of DNA for a database. This has helped investigators solve old cases where only a DNA sample was obtained from the scene. DNA profiling can also be used to identify victims of mass casualty incidents. On the other hand, many convicted people have been released from prison on the basis of DNA techniques, which were not available when a crime had originally been committed.

Bioinformatics: Bioinformatics involves the manipulation, searching, and data mining of biological data, and this includes DNA sequence data. The development of techniques to store and search DNA sequences have led to widely applied advances in computer science, especially string searching algorithms, machine learning and database theory. String searching or matching algorithms, which find an occurrence of a sequence of letters inside a larger sequence of letters, were developed to search for specific sequences of nucleotides. The DNA sequenced may be aligned with other DNA sequences to identify homologous sequences and locate the specific mutations that make them distinct. These techniques, especially multiple sequence alignment, are used in studying phylogenetic relationships and protein function.

Data sets representing entire genomes' worth of DNA sequences, such as those produced by the Human Genome Project, are difficult to use without the annotations that identify the locations of genes and regulatory elements on each chromosome.

Regions of DNA sequence that have the characteristic patterns associated with protein- or RNA-coding genes can be identified by gene finding algorithms, which allow researchers to predict the presence of particular gene products and their possible functions in an organism even before they have been isolated experimentally. Entire genomes may also be compared which can shed light on the evolutionary history of particular organism and permit the examination of complex evolutionary events.

DNA Nanotechnology: DNA nanotechnology uses the unique molecular recognition properties of DNA and other nucleic acids to create self-assembling branched DNA complexes with useful properties. DNA is thus used as a structural material rather than as a carrier of biological information. This has led to the creation of two-dimensional periodic lattices (both tile-based as well as using the "DNA origami" method) as well as three-dimensional structures in the shapes of polyhedra. Nanomechanical devices and algorithmic self-assembly have also been demonstrated, and these DNA structures have been used to template the arrangement of other molecules such as gold nanoparticles and streptavidin proteins.

History and Anthropology: Because DNA collects mutations over time, which are then inherited, it contains historical information, and, by comparing DNA sequences, geneticists can infer the evolutionary history of organisms, their phylogeny. This field of phylogenetics is a powerful tool in evolutionary biology. If DNA sequences within a species are compared, population geneticists can learn the history of particular populations. This can be used in studies ranging from ecological genetics to anthropology; For example, DNA evidence is being used to try to identify the Ten Lost Tribes of Israel. DNA has also been used to look at modern family relationships, such as establishing family relationships between the descendants of Sally Hemings and Thomas Jefferson. This usage is closely related to the use of DNA in criminal investigations detailed above. Indeed, some criminal investigations have been solved when DNA from crime scenes has matched relatives of the guilty individual.

History of DNA Research

DNA was first isolated by the Swiss physician Friedrich Miescher who, in 1869, discovered a microscopic substance in the pus of discarded surgical bandages. As it resided in the nuclei of cells, he called it "nuclein". In 1919, Phoebus Levene identified the base, sugar and phosphate nucleotide unit. Levene suggested that DNA consisted of a string of nucleotide units linked together through the phosphate groups. However, Levene thought the chain was short and the bases repeated in a fixed order. In 1937 William Astbury produced the first X-ray diffraction patterns that showed that DNA had a regular structure.

In 1928, Frederick Griffith discovered that traits of the "smooth" form of the *Pneumococcus* could be transferred to the "rough" form of the same bacteria by mixing killed "smooth" bacteria with the live "rough" form. This system provided the first clear suggestion that DNA carries genetic information—the Avery–MacLeod–McCarty experiment—when Oswald Avery, along with coworkers Colin MacLeod and Maclyn McCarty, identified DNA as the transforming principle in 1943. DNA's role in heredity was confirmed in 1952, when Alfred Hershey and Martha Chase in the Hershey–Chase experiment showed that DNA is the genetic material of the T2 phage.

In 1953, James D. Watson and Francis Crick suggested what is now accepted as the first correct double-helix model of DNA structure in the journal *Nature*. Their double-helix, molecular model of DNA was then based on a single X-ray diffraction image (labeled as "Photo 51") taken by Rosalind Franklin and Raymond Gosling in May 1952, as well as the information that the DNA bases are paired — also obtained through private communications from Erwin Chargaff in the previous years. Chargaff's rules played a very important role in establishing double-helix configurations for B-DNA as well as A-DNA.

Experimental evidence supporting the Watson and Crick model were published in a series of five articles in the same issue of *Nature*. Of these, Franklin and Gosling's paper was the first publication of their own X-ray diffraction data and original analysis method that partially supported the Watson and Crick model; this issue also contained an article on DNA structure by Maurice Wilkins and two of his colleagues, whose analysis and *in vivo* B-DNA X-ray patterns also supported the presence *in vivo* of the double-helical DNA configurations as proposed by Crick and Watson for their double-helix molecular model of DNA in the previous two pages of *Nature*. In 1962, after Franklin's death, Watson, Crick, and Wilkins jointly received the Nobel Prize in Physiology or Medicine. However, Nobel rules of the time allowed only living recipients, but a vigorous debate continues on who should receive credit for the discovery.

In an influential presentation in 1957, Crick laid out the central dogma of molecular biology, which foretold the relationship between DNA, RNA, and proteins, and articulated the "adaptor hypothesis". Final confirmation of the replication mechanism that was implied by the double-helical structure followed in 1958 through the Meselson–Stahl experiment. Further work by Crick and coworkers showed that the genetic code was based on non-overlapping triplets of bases, called codons, allowing Har Gobind Khorana, Robert W. Holley and Marshall Warren Nirenberg to decipher the genetic code. These findings represent the birth of molecular biology.

RNA

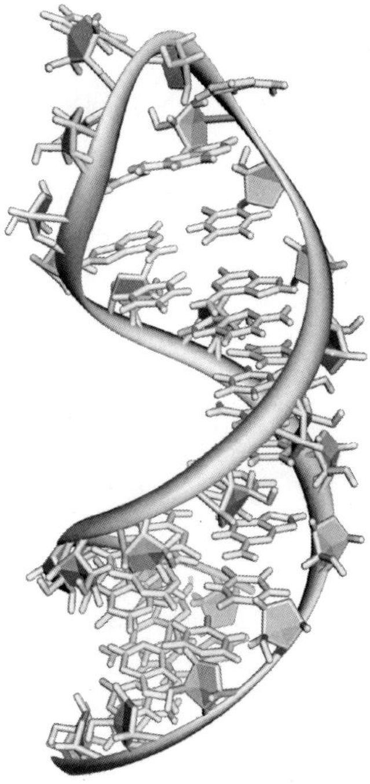

Figure: *A hairpin loop from a pre-mRNA. Highlighted are the nucleobases and the ribose-phosphate backbone*

Ribonucleic acid (RNA) is one of the three major macromolecules (along with DNA and proteins) that are essential for all known forms of life. Like DNA, RNA is made up of a long chain of components called nucleotides. Each nucleotide consists of a nucleobase (sometimes called a nitrogenous base), a ribose sugar, and a phosphate group. The sequence of nucleotides allows RNA to encode genetic information. For example, some viruses use RNA instead of DNA as their genetic material, and all organisms use messenger RNA (mRNA) to carry the genetic information that directs the synthesis of proteins.

Like proteins, some RNA molecules play an active role in cells by catalyzing biological reactions, controlling gene expression, or sensing and communicating responses to cellular signals. One of these active processes is protein synthesis, a universal function whereby mRNA molecules direct the assembly of proteins on ribosomes. This process uses transfer RNA (tRNA) molecules to deliver amino acids to the ribosome, where ribosomal RNA (rRNA) links amino acids together to form proteins.

The chemical structure of RNA is very similar to that of DNA, with two differences—(a) RNA contains the sugar ribose while DNA contains the slightly different sugar deoxyribose (a type of ribose that lacks one oxygen atom), and (b) RNA has the nucleobase uracil while DNA contains thymine (uracil and thymine have similar base-pairing properties).

Unlike DNA, most RNA molecules are single-stranded. Single-stranded RNA molecules adopt very complex three-dimensional structures, since they are not restricted to the repetitive double-helical form of double-stranded DNA. RNA is made within living cells by RNA polymerases, enzymes that act to copy a DNA or RNA template into a new RNA strand through processes known as transcription or RNA replication, respectively.

Comparison with DNA

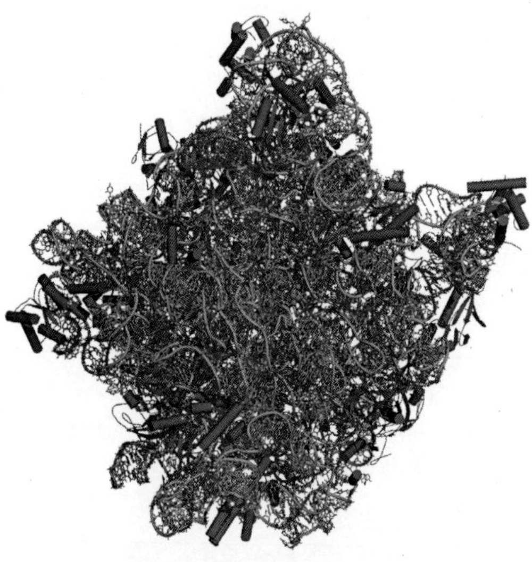

Figure: *Three-dimensional representation of the 50S ribosomal subunit. RNA is in ochre, protein in blue. The active site is in the middle*

RNA and DNA are both nucleic acids, but differ in three main ways. First, unlike DNA, which is, in general, double-stranded, RNA is a single-stranded molecule in many of its biological roles and has a much shorter chain of nucleotides. Second, while DNA contains *deoxyribose*, RNA contains *ribose* (in deoxyribose there is no hydroxyl group attached to the pentose ring in the 2' position). These hydroxyl groups make RNA less stable than DNA because it is more prone to hydrolysis. Third, the complementary base to adenine is not thymine, as it is in DNA, but rather uracil, which is an unmethylated form of thymine.

Like DNA, most biologically active RNAs, including mRNA, tRNA, rRNA, snRNAs, and other non-coding RNAs, contain self-complementary sequences that allow parts of the RNA to fold and pair with itself to form double helices. Structural analysis of these RNAs has revealed that they are highly structured. Unlike DNA, their structures do not consist of long double helices but rather collections of short helices packed together into structures akin to proteins. In this fashion, RNAs can achieve chemical catalysis, like enzymes. For instance, determination of the structure of the ribosome—an enzyme that catalyzes peptide bond formation—revealed that its active site is composed entirely of RNA.

Structure

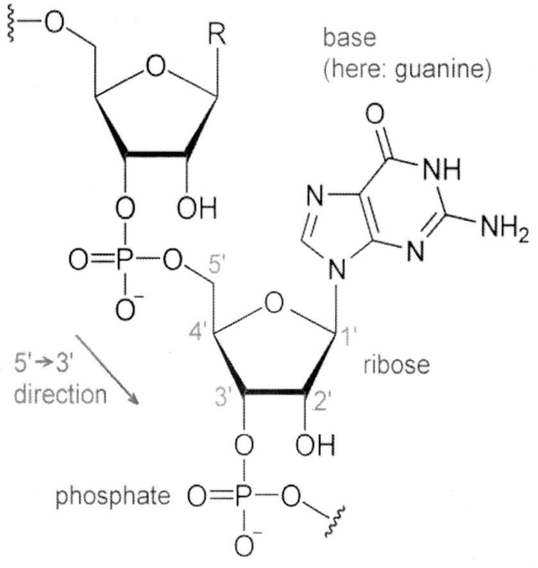

Figure: *Chemical structure of RNA*

Each nucleotide in RNA contains a ribose sugar, with carbons numbered 1' through 5'. A base is attached to the 1' position, in general, adenine (A), cytosine (C), guanine (G), or uracil (U). Adenine and guanine are purines, cytosine, and uracil are pyrimidines. A phosphate group is attached to the 3' position of one ribose and the 5' position of the next. The phosphate groups have a negative charge each at physiological pH, making RNA a charged molecule (polyanion). The bases may form hydrogen bonds between cytosine and guanine, between adenine and uracil and between guanine and uracil. However, other interactions are possible, such as a group of adenine bases binding to each other in a bulge, or the GNRA tetraloop that has a guanine–adenine base-pair.

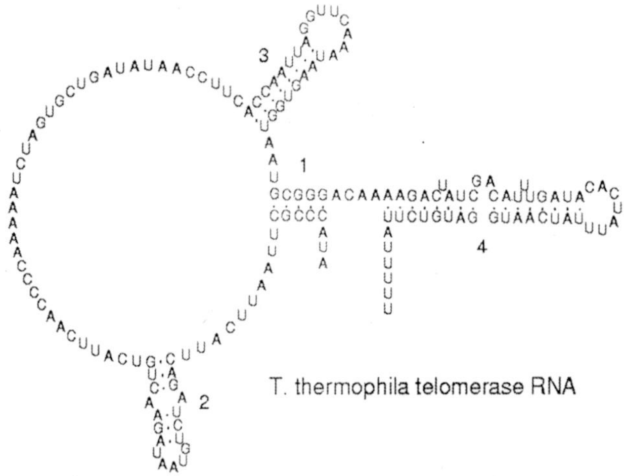

Figure *Secondary structure of a telomerase RNA.*

An important structural feature of RNA that distinguishes it from DNA is the presence of a hydroxyl group at the 2' position of the ribose sugar. The presence of this functional group causes the helix to adopt the A-form geometry rather than the B-form most commonly observed in DNA. This results in a very deep and narrow major groove and a shallow and wide minor groove. A second consequence of the presence of the 2'-hydroxyl group is that in conformationally flexible regions of an RNA molecule (that is, not involved in formation of a double helix), it can chemically attack the adjacent phosphodiester bond to cleave the backbone. RNA is transcribed with only four bases (adenine, cytosine, guanine and uracil), but these bases and attached sugars can be modified in numerous ways as the RNAs mature. Pseudouridine (Ø), in which the linkage between uracil and ribose is changed from a C–N bond to a C–C bond, and ribothymidine (T) are found in various places (the most notable ones being in the TØC loop of tRNA). Another notable modified base is hypoxanthine, a deaminated adenine base whose nucleoside is called inosine (I). Inosine plays a key role in the wobble hypothesis of the genetic code.

There are nearly 100 other naturally occurring modified nucleosides, of which pseudouridine and nucleosides with 2'-O-methylribose are the most common. The specific roles of many of these modifications in RNA are not fully understood. However, it is notable that, in ribosomal RNA, many of the post-transcriptional modifications occur in highly functional regions, such as the peptidyl transferase centre and the subunit interface, implying that they are important for normal function. The functional form of single stranded RNA molecules, just like proteins, frequently requires a specific tertiary structure. The scaffold for this structure is provided by secondary structural elements that are hydrogen bonds within the molecule. This leads to several recognizable "domains" of secondary structure like hairpin loops, bulges, and internal loops. Since RNA is charged, metal ions such as Mg2+ are needed to stabilise many secondary and tertiary structures.

Synthesis

Synthesis of RNA is usually catalyzed by an enzyme—RNA polymerase—using DNA as a template, a process known as transcription. Initiation of transcription begins with the binding of the enzyme to a promoter sequence in the DNA (usually found "upstream" of a gene). The DNA double helix is unwound by the helicase activity of the enzyme. The enzyme then progresses along the template strand in the 3' to 5' direction, synthesizing a complementary RNA molecule with elongation occurring in the 5' to 3' direction. The DNA sequence also dictates where termination of RNA synthesis will occur. RNAs are often modified by enzymes after transcription. For example, a poly(A) tail and a 5' cap are added to eukaryotic pre-mRNA and introns are removed by the spliceosome. There are also a number of RNA-dependent RNA polymerases that use RNA as their template for synthesis of a new strand of RNA. For instance, a number of RNA viruses (such as poliovirus) use this type of enzyme to replicate their genetic material. Also, RNA-dependent RNA polymerase is part of the RNA interference pathway in many organisms.

Types of RNA

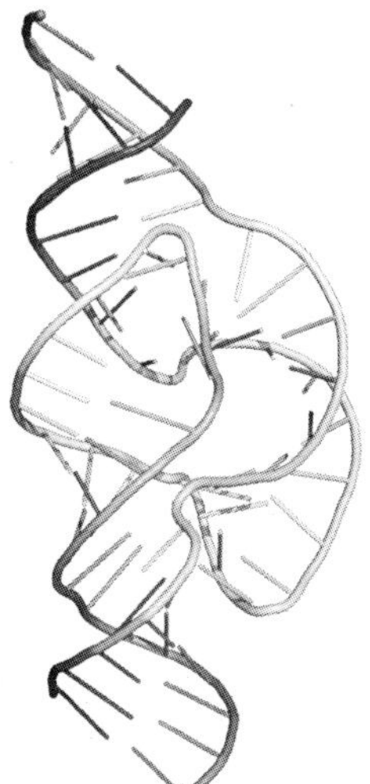

Figure 14: *Structure of a hammerhead ribozyme, a ribozyme that cuts RNA*

Messenger RNA (mRNA) is the RNA that carries information from DNA to the ribosome, the sites of protein synthesis (translation) in the cell. The coding sequence of the mRNA determines the amino acid sequence in the protein that is produced. Many RNAs do not code for protein however (about 97% of the transcriptional output is non-protein-coding in eukaryotes). These so-called non-coding RNAs ("ncRNA") can be encoded by their own genes (RNA genes), but can also derive from mRNA introns. The most prominent examples of non-coding RNAs are transfer RNA (tRNA) and ribosomal RNA (rRNA), both of which are involved in the process of translation. There are also non-coding RNAs involved in gene regulation, RNA processing and other roles. Certain RNAs are able to catalyse chemical reactions such as cutting and ligating other RNA molecules, and the catalysis of peptide bond formation in the ribosome; these are known as ribozymes.

In Translation: Messenger RNA (mRNA) carries information about a protein sequence to the ribosomes, the protein synthesis factories in the cell. It is coded so that every three nucleotides (a codon) correspond to one amino acid. In eukaryotic cells, once precursor mRNA (pre-mRNA) has been transcribed from DNA, it is processed to mature mRNA. This removes its introns—non-coding sections of the pre-mRNA. The mRNA is then exported from the nucleus to the cytoplasm, where it is bound to ribosomes and translated into its corresponding protein form with the help of tRNA. In prokaryotic cells, which do not have nucleus and cytoplasm compartments, mRNA can bind to ribosomes while it is being transcribed from DNA. After a certain amount of time the message degrades into its component nucleotides with the assistance of ribonucleases. Transfer RNA (tRNA) is a small RNA chain of about 80 nucleotides that transfers a specific amino acid to a growing polypeptide chain at the ribosomal site of protein synthesis during translation. It has sites for amino acid attachment and an anticodon region for codon recognition that binds to a specific sequence on the messenger RNA chain through hydrogen bonding. Ribosomal RNA (rRNA) is the catalytic component of the ribosomes. Eukaryotic ribosomes contain four different rRNA molecules: 18S, 5.8S, 28S and 5S rRNA. Three of the rRNA molecules are synthesized in the nucleolus, and one is synthesized elsewhere. In the cytoplasm, ribosomal RNA and protein combine to form a nucleoprotein called a ribosome. The ribosome binds mRNA and carries out protein synthesis. Several ribosomes may be attached to a single mRNA at any time. rRNA is extremely abundant and makes up 80% of the 10 mg/ml RNA found in a typical eukaryotic cytoplasm.

Transfer-messenger RNA (tmRNA) is found in many bacteria and plastids. It tags proteins encoded by mRNAs that lack stop codons for degradation and prevents the ribosome from stalling.

Regulatory RNAs: Several types of RNA can downregulate gene expression by being complementary to a part of an mRNA or a gene's DNA. MicroRNAs (miRNA; 21-22 nt) are found in eukaryotes and act through RNA interference (RNAi), where an effector complex of miRNA and enzymes can break down mRNA to which the miRNA is complementary, block the mRNA from being translated, or accelerate its degradation. While small interfering RNAs (siRNA; 20-25 nt) are often produced by breakdown of viral RNA, there are also endogenous sources of siRNAs. siRNAs act through RNA interference in a fashion similar to miRNAs. Some miRNAs and siRNAs can cause genes they target to be methylated, thereby decreasing or increasing transcription of those genes. Animals have Piwi-interacting RNAs (piRNA; 29-30 nt) which are active in germline cells and are thought to be a defence against transposons and play a role in gametogenesis.

Many prokaryotes have CRISPR RNAs, a regulatory system similar to RNA interference. Antisense RNAs are widespread; most downregulate a gene, but a few are activators of transcription. One way antisense RNA can act is by binding to an mRNA, forming double-stranded RNA that is enzymatically degraded. There are many long noncoding RNAs that regulate genes in eukaryotes, one such RNA is Xist, which coats one X chromosome in female mammals and inactivates it.

An mRNA may contain regulatory elements itself, such as riboswitches, in the 5' untranslated region or 3' untranslated region; these cis-regulatory elements regulate the activity of that mRNA. The untranslated regions can also contain elements that regulate other genes.

In RNA Processing: Many RNAs are involved in modifying other RNAs. Introns are spliced out of pre-mRNA by spliceosomes, which contain several small nuclear RNAs (snRNA), or the introns can be ribozymes that are spliced by themselves. RNA can also be altered by having its nucleotides modified to other nucleotides than A, C, G and U.

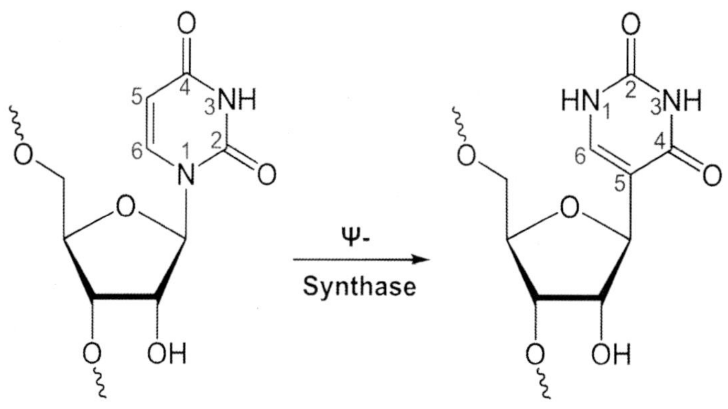

Figure: *Uridine to pseudouridine is a common RNA modification.*

In eukaryotes, modifications of RNA nucleotides are generally directed by small nucleolar RNAs (snoRNA; 60-300 nt), found in the nucleolus and cajal bodies. snoRNAs associate with enzymes and guide them to a spot on an RNA by basepairing to that RNA. These enzymes then perform the nucleotide modification. rRNAs and tRNAs are extensively modified, but snRNAs and mRNAs can also be the target of base modification.

RNA Genomes: Like DNA, RNA can carry genetic information. RNA viruses have genomes composed of RNA, and a variety of proteins encoded by that genome. The viral genome is replicated by some of those proteins, while other proteins protect the genome as the virus particle moves to a new host cell. Viroids are another group of pathogens, but they consist only of RNA, do not encode any protein and are replicated by a host plant cell's polymerase.

In Reverse Transcription: Reverse transcribing viruses replicate their genomes by reverse transcribing DNA copies from their RNA; these DNA copies are then transcribed to new RNA. Retrotransposons also spread by copying DNA and RNA from one another, and telomerase contains an RNA that is used as template for building the ends of eukaryotic chromosomes.

Double-Stranded RNA: Double-stranded RNA (dsRNA) is RNA with two complementary strands, similar to the DNA found in all cells. dsRNA forms the genetic material of some viruses (double-stranded RNA viruses). Double-stranded RNA such as viral RNA or siRNA can trigger RNA interference in eukaryotes, as well as interferon response in vertebrates.

Key Discoveries in RNA Biology

Research on RNA has led to many important biological discoveries and numerous Nobel Prizes. Nucleic acids were discovered in 1868 by Friedrich Miescher, who called the material 'nuclein' since it was found in the nucleus. It was later discovered that prokaryotic cells, which do not have a nucleus, also contain nucleic acids. The role of RNA in protein synthesis was suspected already in 1939. Severo Ochoa won the 1959 Nobel Prize in Medicine (shared with Arthur Kornberg) after he discovered an enzyme that can synthesize RNA in the laboratory. Ironically, the enzyme discovered by Ochoa (polynucleotide phosphorylase) was later shown to be responsible for RNA degradation, not RNA synthesis. The sequence of the 77 nucleotides of a yeast tRNA was found by Robert W. Holley in 1965, winning Holley the 1968 Nobel Prize in Medicine (shared with Har Gobind Khorana and Marshall Nirenberg). In 1967, Carl Woese hypothesized that RNA might be catalytic and suggested that the earliest forms of life (self-replicating molecules) could have relied on RNA both to carry genetic information and to catalyze biochemical reactions—an RNA world.

During the early 1970s, retroviruses and reverse transcriptase were discovered, showing for the first time that enzymes could copy RNA into DNA (the opposite of the usual route for transmission of genetic information). For this work, David Baltimore, Renato Dulbecco and Howard Temin were awarded a Nobel Prize in 1975. In 1976, Walter Fiers and his team determined the first complete nucleotide sequence of an RNA virus genome, that of bacteriophage MS2.

In 1977, introns and RNA splicing were discovered in both mammalian viruses and in cellular genes, resulting in a 1993 Nobel to Philip Sharp and Richard Roberts.

Catalytic RNA molecules (ribozymes) were discovered in the early 1980s, leading to a 1989 Nobel award to Thomas Cech and Sidney Altman. In 1990 it was found in petunia that introduced genes can silence similar genes of the plant's own, now known to be a result of RNA interference.

At about the same time, 22 nt long RNAs, now called microRNAs, were found to have a role in the development of *C. elegans*. Studies on RNA interference gleaned a Nobel Prize for Andrew Fire and Craig Mello in 2006, and another Nobel was awarded for studies on transcription of RNA to Roger Kornberg in the same year. The discovery of gene regulatory RNAs has led to attempts to develop drugs made of RNA, such as siRNA, to silence genes.

Nucleotide

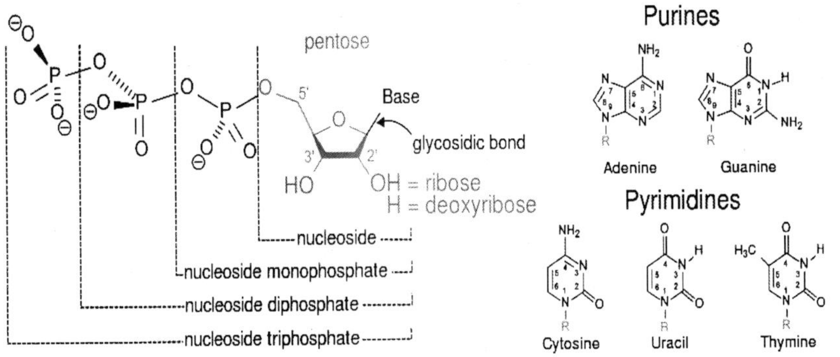

Figure: *Structural elements of the most common nucleotides*

Nucleotides are molecules that, when joined together, make up the structural units of RNA and DNA. In addition, nucleotides play central roles in metabolism.

In that capacity, they serve as sources of chemical energy (adenosine triphosphate and guanosine triphosphate), participate in cellular signaling (cyclic guanosine monophosphate and cyclic adenosine monophosphate), and are incorporated into important cofactors of enzymatic reactions (coenzyme A, flavin adenine dinucleotide, flavin mononucleotide, and nicotinamide adenine dinucleotide phosphate).

Nucleotide Structure

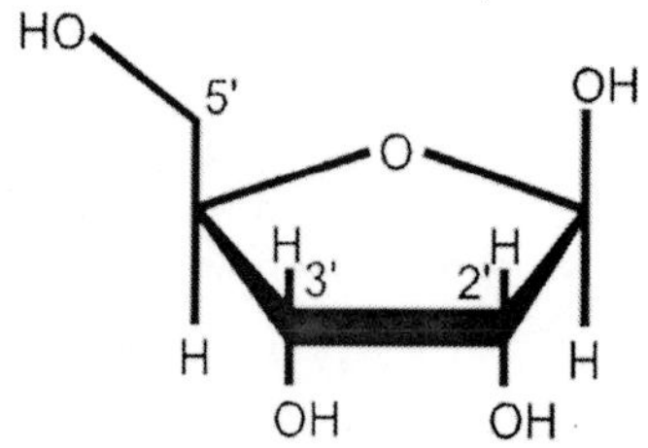

Figure: *Ribose structure indicating numbering of carbon atoms*

A nucleotide is composed of a nucleobase (nitrogenous base), a five-carbon sugar (either ribose or 2'-deoxyribose), and one to three phosphate groups. Together, the nucleobase and sugar comprise a nucleoside. The phosphate groups form bonds with either the 2, 3, or 5-carbon of the sugar, with the 5-carbon site most common. Cyclic nucleotides form when the phosphate group is bound to two of the sugar's hydroxyl groups. Ribonucleotides are nucleotides where the sugar is ribose, and deoxyribonucleotides contain the sugar deoxyribose. Nucleotides can contain either a purine or a pyrimidine base. Nucleic acids are polymeric macromolecules made from nucleotide monomers. In DNA, the purine bases are adenine and guanine, while the pyrimidines are thymine and cytosine. RNA uses uracil in place of thymine. Adenine always pairs with thymine by 2 hydrogen bonds, while guanine pairs with cytosine through 3 hydrogen bonds, each due to their unique structures.

Synthesis

Nucleotides can be synthesized by a variety of means both in vitro and in vivo. In vivo, nucleotides can be synthesised de novo or recycled through salvage pathways. Nucleotides undergo breakdown such that useful parts can be reused in synthesis reactions to create new nucleotides. In vitro, protecting groups may be used during laboratory production of nucleotides. A purified nucleoside is protected to create a phosphoramidite, which can then be used to obtain analogues not found in nature and/or to synthesize an oligonucleotide.

Pyrimidine Ribonucleotides: Pyrimidine nucleotide synthesis starts with the formation of carbamoyl phosphate from glutamine and CO_2. The cyclisation reaction between carbamoyl phosphate reacts with aspartate, yielding orotate in subsequent steps. Orotate reacts with 5-phosphoribosyl á-diphosphate (PRPP), yielding orotidine monophosphate (OMP), which is decarboxylated to form uridine monophosphate (UMP). It is from UMP that other pyrimidine nucleotides are derived. UMP is phosphorylated to uridine triphosphate (UTP) via two sequential reactions with ATP. Cytidine monophosphate (CMP) is derived from conversion of UTP to cytidine triphosphate (CTP) with subsequent loss of two phosphates.

Purine Ribonucleotides: The atoms which are used to build the purine nucleotides come from a variety of sources:

- The biosynthetic origins of purine ring atoms
- N1 arises from the amine group of Asp
- C2 and C8 originate from formate
- N3 and N9 are contributed by the amide group of Gln
- C4, C5 and N7 are derived from Gly
- C6 comes from HCO_3^- (CO_2)

The de novo synthesis of purine nucleotides by which these precursors are incorporated into the purine ring proceeds by a 10-step pathway to the branch-point intermediate IMP, the nucleotide of the base hypoxanthine. AMP and GMP are subsequently synthesized from this intermediate via separate, two-step pathways. Thus, purine moieties are initially formed as part of the ribonucleotides rather than as free bases.

Six enzymes take part in IMP synthesis. Three of them are multifunctional:

- GART (reactions 2, 3, and 5)
- PAICS (reactions 6, and 7)
- ATIC (reactions 9, and 10)

Reaction 1. The pathway starts with the formation of PRPP. PRPS1 is the enzyme that activates R5P, which is formed primarily by the pentose phosphate pathway, to PRPP by reacting it with ATP. The reaction is unusual in that a pyrophosphoryl group is directly transferred from ATP to C1 of R5P and that the product has the á configuration about C1. This reaction is also shared with the pathways for the synthesis of the pyrimidine nucleotides, Trp, and His. As a result of being on (a) such (a) major metabolic crossroad and the use of energy, this reaction is highly regulated.

Reaction 2. In the first reaction unique to purine nucleotide biosynthesis, PPAT catalyzes the displacement of PRPP's pyrophosphate group (PP_i) by Gln's amide nitrogen. The reaction occurs with the inversion of configuration about ribose C1, thereby forming â-5-phosphorybosylamine (5-PRA) and establishing the anomeric form of the future nucleotide. This reaction, which is driven to completion by the subsequent hydrolysis of the released PP_i, is the pathway's flux-generating step and is therefore regulated, too.

Length Unit

Nucleotide (abbreviated nt) is a common length unit for single-stranded RNA, similar to how base pair is a length unit for double-stranded DNA.

Gluconeogenesis

Gluconeogenesis (abbreviated GNG) is a metabolic pathway that results in the generation of glucose from non-carbohydrate carbon substrates such as lactate, glycerol, and glucogenic amino acids. It is one of the two main mechanisms humans and many other animals use to keep blood glucose levels from dropping too low (hypoglycemia). The other means of maintaining blood glucose levels is through the degradation of glycogen (glycogenolysis). Gluconeogenesis is a ubiquitous process, present in plants, animals, fungi, bacteria, and other microorganisms. In animals, gluconeogenesis takes place mainly in the liver and, to a lesser extent, in the cortex of kidneys. This process occurs during periods of fasting, starvation, low-carbohydrate diets, or intense exercise and is highly endergonic. For example, the pathway leading from phosphoenolpyruvate to glucose-6-phosphate requires 6 molecules of ATP. Gluconeogenesis is often associated with ketosis. Gluconeogenesis is also a target of therapy for type II diabetes, such as metformin, which inhibits glucose formation and stimulates glucose uptake by cells.

Entering the Pathway

Lactate is transported back to the liver where it is converted into pyruvate by the Cori cycle using the enzyme lactate dehydrogenase. Pyruvate, the first designated substrate of the gluconeogenic pathway, can then be used to generate glucose. All citric acid cycle intermediates, through conversion to oxaloacetate, amino acids other than lysine or leucine, and glycerol can also function as substrates for gluconeogenesis. Transamination or deamination of amino acids facilitates entering of their carbon skeleton into the cycle directly (as pyruvate or oxaloacetate), or indirectly via the citric acid cycle.

Whether fatty acids can be converted into glucose in animals has been a longstanding question in biochemistry. It is known that odd-chain fatty acids can be oxidized to yield propionyl CoA, a precursor for succinyl CoA, which can be converted to pyruvate and enter into gluconeogenesis. In plants, to be specific, in seedlings, the glyoxylate cycle can be used to convert fatty acids (acetate) into the primary carbon source of the organism. The glyoxylate cycle produces four-carbon dicarboxylic acids that can enter gluconeogenesis.

In 1995, researchers identified the glyoxylate cycle in nematodes. In addition, the glyoxylate enzymes malate synthase and isocitrate lyase have been found in animal tissues. Genes coding for malate synthase gene have been identified in other (metazoans) including arthropods, echinoderms, and even some vertebrates. Mammals found to possess these genes include monotremes (platypus) and marsupials (opossum) but not placental mammals. Genes for isocitrate lyase are found only in nematodes, in which, it is apparent, they originated in horizontal gene transfer from bacteria.

The existence of glyoxylate cycles in humans has not been established, and it is widely held that fatty acids cannot be converted to glucose in humans directly. However, carbon-14 has been shown to end up in glucose when it is supplied in fatty acids. Despite these findings, it is considered unlikely that the 2-carbon acetyl-CoA derived from the oxidation of fatty acids would produce a net yield of glucose via the citric acid cycle. However, it is possible that, with additional sources of carbon via other pathways, glucose could be synthesized from acetyl-CoA. In fact, it is known that Ketone bodies, â-hydroxybutyrate in particular, can be converted to glucose at least in small amounts (â-hydroxybutyrate to acetoacetate to acetone to propanediol to pyruvate to glucose). Glycerol, which is a part of the triacylglycerol molecule, can be used in gluconeogenesis.

Location

In humans, gluconeogenesis is restricted to the liver and to a lesser extent the kidney. In all species, the formation of oxaloacetate from pyruvate and TCA cycle intermediates is restricted to the mitochondrion, and the enzymes that convert PEP to glucose are found in the cytosol. The location of the enzyme that links these two parts of gluconeogenesis by converting oxaloacetate to PEP, PEP carboxykinase, is variable by species: it can be found entirely within the mitochondria, entirely within the cytosol, or dispersed evenly between the two, as it is in humans.

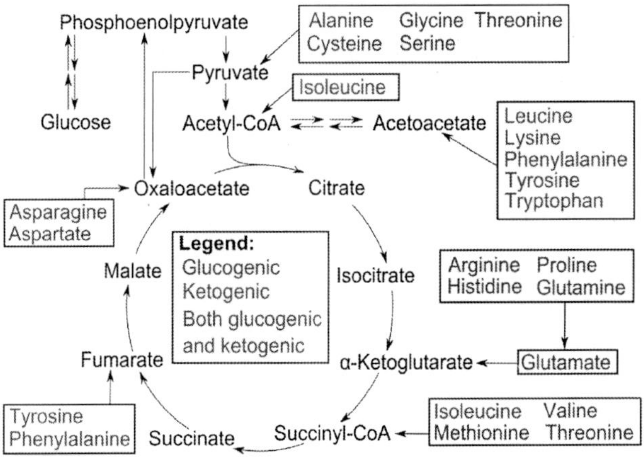

Figure: *Catabolism of proteinogenic amino acids. Amino acids are classified according the abilities of their products to enter gluconeogenesis: Glucogenic amino acids have this ability; Ketogenic amino acids do not. These products may still be used for ketogenesis or lipid synthesis; Some amino acids are catabolized into both glucogenic and ketogenic products.*

Transport of PEP across the mitochondrial membrane is accomplished by dedicated transport proteins; however no such proteins exist for oxaloacetate. Therefore species that lack intra-mitochondrial PEP, oxaloacetate must be converted into malate or asparate, exported from the mitochondrion, and converted back into oxaloacetate in order to allow gluconeogenesis to continue.

Pathway

Gluconeogenesis is a pathway consisting of eleven enzyme-catalyzed reactions. The pathway can begin in the mitochondria or cytoplasm, depending on the substrate being used. Many of the reactions are the reversible steps found in glycolysis.

- Gluconeogenesis begins in the mitochondria with the formation of oxaloacetate through carboxylation of pyruvate. This reaction also requires one molecule of ATP, and is catalyzed by pyruvate carboxylase. This enzyme is stimulated by high levels of acetyl-CoA (produced in â-oxidation in the liver) and inhibited by high levels of ADP.
- Oxaloacetate is reduced to malate using NADH, a step required for transport out of the mitochondria.
- Malate is oxidized to oxaloacetate using NAD^+ in the cytoplasm, where the remaining steps of gluconeogenesis occur.
- Oxaloacetate is decarboxylated and phosphorylated to produce phosphoenolpyruvate by phosphoenolpyruvate carboxykinase. One molecule of GTP is hydrolyzed to GDP during this reaction.
- The next steps in the reaction are the same as reversed glycolysis. However, fructose-1,6-bisphosphatase converts fructose-1,6-bisphosphate to fructose 6-phosphate, requiring one water molecule and releasing one phosphate. This is also the rate-limiting step of gluconeogenesis.
- Glucose-6-phosphate is formed from fructose 6-phosphate by phosphoglucoisomerase. Glucose-6-phosphate can be used in other metabolic pathways or dephosphorylated to free glucose. Whereas free glucose can easily diffuse in and out of the cell, the phosphorylated form (glucose-6-phosphate) is locked in the cell, a mechanism by which intracellular glucose levels are controlled by cells.
- The final reaction of gluconeogenesis, the formation of glucose, occurs in the lumen of the endoplasmic reticulum, where glucose-6-phosphate is hydrolyzed by glucose-6-phosphatase to produce glucose. Glucose is shuttled into the cytosol by glucose transporters located in the membrane of the endoplasmic reticulum.

Regulation

While most steps in gluconeogenesis are the reverse of those found in glycolysis, three regulated and strongly exergonic reactions are replaced with more kinetically favourable reactions. Hexokinase/glucokinase, phosphofructokinase, and pyruvate kinase enzymes of glycolysis are replaced with glucose-6-phosphatase, fructose-1,6-bisphosphatase, and PEP carboxykinase. This system of reciprocal control allow glycolysis and gluconeogenesis to inhibit each other and prevent the formation of a futile cycle. The majority of the enzymes responsible for gluconeogenesis are found in the cytoplasm; the exceptions are mitochondrial pyruvate carboxylase and, in animals, phosphoenolpyruvate carboxykinase. The latter exists as an isozyme located in both the mitochondrion and the cytosol. The rate of gluconeogenesis is ultimately controlled by the action of a key enzyme, fructose-1,6-bisphosphatase, which is also regulated through signal tranduction by cAMP and its phosphorylation.

Most factors that regulate the activity of the gluconeogenesis pathway do so by inhibiting the activity or expression of key enzymes. However, both acetyl CoA and citrate activate gluconeogenesis enzymes (pyruvate carboxylase and fructose-1,6-bisphosphatase, respectively). Due to the reciprocal control of the cycle, acetyl-CoA and citrate also have inhibitory roles in the activity of pyruvate kinase. Global control of gluconeogenesis is mediated by glucagon (*released when blood glucose is low*); it triggers phosphorylation of enzymes and regulatory proteins by Protein Kinase A (a cyclic AMP regulated kinase) resulting in inhibition of glycolysis and stimulation of gluconeogenesis, thus *bringing blood glucose levels up.*

7

The Genetics System

The genes are much too small to be seen individually, but in certain stages many chromosomes show lengthwise patterns of knobs or bands that indicate their ubdivision into genes. In particular, this happens during the early stages of meiosis. A pair of rye chromosomes at the beginning of meiosis. You notice the pattern of darkly stained knobs and faintly stained connecting strands. The pattern in the two partner chromosomes is the same, and pairing is exceedingly accurate, each knob in one chromosome lying next to the matching knob in its partner. It appears that it is the genes rather than the chroosomes as a whole that direct the pairing process. This ability of the genes to find and attract their exact counterparts is one of their many remarkable properties for which science has not yet found a satisfactory explanation.

Later on in meiosis the pattern will be lost because the chromosomes curl up and contact into the compact bodies. But when the next generation makes germ cells, each chromosome will again show its characteristic pattern, so that a trained cytologist can recognize every one of the several chromosomes of rye by the pattern of its knobs. Giant chromosomes with beautiful and intricate patterns of dar and light bands are found in the salivary glands of the larvae (maggts) of certain flies. Chromosomes in the salivary glands of Drosophila larvae. They are hundreds of times as long as ordinary chromosomes, and each of them has its own characteristic pattern of bands. They are also much thicker than ordinary chromsomes because each consists of a bundle of many strands, all with exactly the same pattern of bands. The way these bundles arise is intresting because it illustrates the extreme accuracy of chromosome replication. In the young larva, the chromosomes in the salivary glands are still thin.

As the larva grows, the salivary glands grow too, but they do so in a particular way.

The cells, instead of making more cells by division, grow bigger and bigger without division. Their nuclei, too, grow bigger and bigger. Inside the nuclei, the chromosomes make replicas of themselves as they do in preparation for mitosis but, as there is no mitosis, the new chromosomes do not separate from the old ones. Instead, new and old chromosomes remain closely attached to each other and each, in turn, makes more replicas. This goes on repeatedly, and so accurately does each band make another band like itself that the whole bundle appears as one single thick banded thread. In addition, the partner cromosomes in the salivary glands are closely paired as in meiosis, so that each apparent chromosome represents one pair, and the nuclei of the salivary glands of Drosophila seem to have four chromosomes instead of the eight that are visible in ordinary cells).

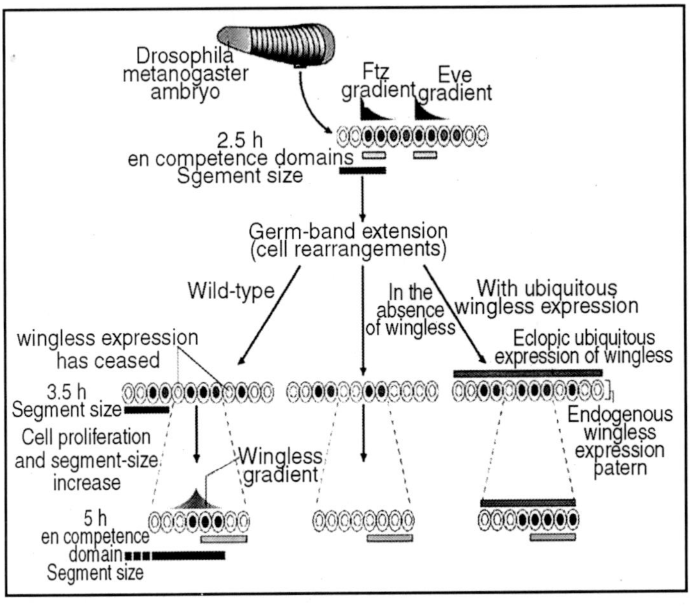

Figure: *Wing Gene in Drosophila*

The pattern of bands corresponds to the pattern of underlying genes. This can often be seen in cases where a chromosome has lost a small piece through radiation or some other injury. The accuracy of chromosome replication is such that all descendants of the damaged chromosome lack exactly the same piece, and when such a damaged chromosome gets into a gamete, for instance an ovum, individuals are formed which lack the same chromosome piece in all their cells. The left part of a normal wing of Drosophila and above it, and at a very much higher mgnification, a little section of one of the normal salivary gland chromosomes. The arrow points to a band which encloses a gene for wing shape. The right part of the same chromosome section in afly which had received a damaged chromosome from its X-rayed father or mother.

If you recall that the salivary gland chromosomes consist of two chromosome bundles formed by the two partner chromosomes, you will unerstand why only half a band is missing at the point of the arrow. Nevertheless, the lack of one of the two partner genes has been sufficient to disturb normal development of thewing and to produce the abnormal shape shown in the figure. Flies which have lost this band from both partner chromosomes are so much damaged that they die as early embryos. The left drawing represents a normally growing green maize plant and underneath, at much higher magnification, the tips of one particular chromosome pair. Thee is a fairly large knob at the end of these chromosomes. Among the genes which this knob encloses is one whose action is necessary for the production of the green substance chlorophyll that is essential for survival of the plant.

The right picture shows that, when this knob has been lost from both chromosomes, the plant dies early as a yellow and sickly seedling. If you turn once more to the left picture you will notice a striking difference from what happened in the case of the Drosophila wing gene. Whereas this gene could not produce anormal wing without the co-operation of its partner gen, the chlorophyll gene is sufficiently effective even when acting singly; for plants with one normal and one knobless chromosome are quite normal.

This is not a distinction between the genes of Drosophila and maize, or between genes concerned with wing hape or chlorophyll.

It is a difference in the efficiency with which genes carry out their functions; some genes are elfsufficient in this regard, others need the co-operation of their partners for effective action. We shall meet a similar difference when we deal with dominant and recessive genes. Each gene is concerned in the control of a particular developmetal process, for instance the formation of chlorophyll, the development of colour or size or, in higher animals, mental ability. Partner genes control the same process, but they may do so in different ways; thus two partner genes in peas are concerned with the control of seed shape, but one causesthe seeds to be round, while the other makes them wrinkled.

Partner genes with different effects are called alleloorphs or, for short, alleles. The term "allelomorph" signifies "diffrent shape"; according to modern ideas this may be interpreted as different chemical configurations of basically the same gene. All developmental processes are highly complex, and it is therefore not surprising that many different gene pairs are involved in the develpment of, say, an eye or a brain or a flower. Even such a seemingly simple achievement as the formation of a vitamin by a bacterium proceeds in a series of steps and is controlled by a series of genes. In bacteria, genes controlling the same process tend to lie close together, as though this facilitated their co-operation; but in most other organisms this is not found to be so. Genes which are involved in quite different processes may lie cheek by jowl on the same chromosome, while genes controlling closely related processes may lie far apart. The position of a few of the known genes on a particular pair of chromosomes of the fruit fly Drosophila melanogaster.

Only genes concerned with the development of the eyes and the wings are represented; more genes controlling development of these organs are present on the same chromosome pair and on the three others. Some of the represented genes can be seen to act in very different ways; thus wing gene w_1 is concerned with the control of wing shape, while wing gene w_2 helps make the correct pattern of wing venation. For other genes such a distinction is not evident; thus eye genes e_1, e_2, e_3 and e_4 are all engaged in the formation of eye colour. Yet we know from developmenal studies that each of them has a different role to play in this process.

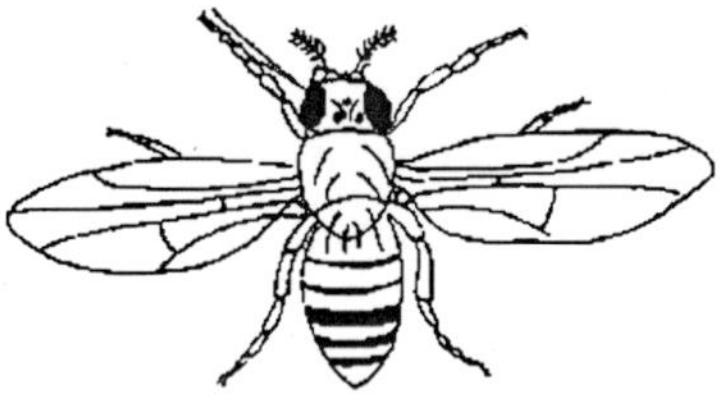

Figure: *The Positions of Some Wing and Eye Genes*

The genes, like the chromosomes, are arranged in pairs; with the chromosomes they segregate into different gametes at meiosis and come together again at fertilization, one member of each pair being derived from the father, one from the mother. The arrangement of the genes on the chromosomes is fixed for each species, and partner genes occupy identical positions on partner chromosomes. Allelomorphs are partner genes with different effect, that is, they are genes whch occupy the same position on the same chromosome pair and control exactly the same developmental process, but they do so in different ways.

Genetic Code

Even more remarkable is that non-overlapping triples of the letters *A, C, G* and *T* which are often reerred to as codons (especially when the triples are part of a gene), may spell out (reading in the 5' to 3' direction) the order of the amino acids that form the long linear molecules known as proteins. It turns out there are 20 amino acids which are the building blocks for proteins. Since there are 64 possible triples (repeated letters within the triples are allowed) using the 4 letters *A, C, G,* and *T*, different triples can represent the same amino acids. There are also 3 triples (*TAA, TAG, TGA*) known as stop codons that do not represent (a common phrase being do not code for) an amino acid, but are involved in signaling a termination to the protein production process. One particular codon (*ATG*), which does represent one of the amino acids, is an indication for the start of the production of a protein. This system for coding proteins has come to be known as the genetic code.

Before the genetic code was fully understood, based on work by Marshall Nirenberg, Heinrich Matthaei, Har Gobind Khorana, and others, various mathematical ideas were invoked to suggest what kind of code might be involved. There is a diagram of the RNA version of the genetic code, where *U* takes the place of *T*.

One approach to finding a gene is to look for a stretch of DNA which starts with ATG (the start codon) and ends with one of the stop or termination codons. This situation in a stretch of DNA is referred to as an open reading frame (ORF). However, not all ORFs correspond to a stretch of DNA which will initiate the production of a protein, and so the location of an ORF is not equivalent to finding a gene. A major problem facing biologists is how to locate those stretches of DNA that are genes in the large amounts of DNA found in a chromosome.

The way that DNA is involved in the production of proteins is not direct. The mechanism involves another helical molecule called RNA, of which there are a variety of types. (Among the types of RNA are messenger RNA, or mRNA, ribosomal RNA, or rRNA, and transfer RNA, or tRNA.) When a protein is to be produced, the DNA separates and a copy of the gene is transcribed (the process is referred to as transcription by molecular biologists) into a strand of RNA (ribonucleic acid). RNA, like DNA, uses an alphabet of 4 nucleotides *A, C, G,* and*U* (for uracil, a pyrimidine like thymine which takes the place of *T*). In a somewhat complex series of steps, a protein is produced. A very rough schematic of how genes create proteins. This schematic is often referred to as the central dogma. The original DNA, which are known as introns but which are not involved in the production of a protein, are separated or snipped out and a resulting string of exons from the original DNA is left. What is left is sometimes referred to as a coding sequence, and it contains the codons (in the letters *A, C, G* and *U*) which will create the protein, by the process usually referred to as translation.

The actual manufacture of the proteins involves structures called ribosomes. Production of proteins does not go on within the nucleus but in the cytoplasm surrounding the nucleus. One can express the codons involved in the manufacture of proteins either in DNA terms (using the letters *A, C, G* and *T*) or in RNA terms (using the letters *A, C, G* and *U*).

Implications

Many interesting quesions can be asked about the genetic code, including why the pattern of redundancies in the code exists. While at one time it was thought that the code was totally universal, this has proved not to be true. For certain bacteria and in other situations the universality of the code breaks down. Cellular life forms are classified according to whether they are prokaryotes or eukaryotes. Prokaryotes are cellular organisms where the genetic material is not a part of a membrane enclosed nucleus, while in eukaryotes the genetic material is contained in a nucleus.

At one time life forms were classified as either bacteria or eukaryotes. A group of bacteria like organisms, found in very hot environments such as undersea vents, were initially classified as bacteria. However, when Archaea's genetic makeup was studied, its members were found to have an amazingly large number of genes that were not known to exist in other organisms. They seemed to be surprisingly different from the bacteria with which they had been formerly classified.

It has become standard in recent years to list the Archaea in a separate category from the bacteria and eukaryotic organisms. Use of phylogentic trees was in part a tool that led to this change. It is tempting to believe that, because human beings are as complex as they are, other species will have less genetic material than humans. However, chromosome number (and the total size of the genome) varies with species in a way that seems to have no rhyme or reason. Thus, chimpanzees have 48 chromosomes, cats 38, dogs 78, rabbits 44, rats 42, turkeys 82, and horses 64. Perhaps when the genomes of many different and varied species are sequenced it will be possible to get insight into these wildly varying numbers.

Mendel's First Law: Segregation

After this excursion into cytology, let us return to genetics. The chief tool of the cytologist is the microscope; it has shown him that the chromosomes, and with them the genes, segregate at meiosis and enter the gametes singly. The chief tool of the geneticist is the crossing of different strains, and we are now going to consider crosses which demonstrate the segregation of the genes at meiosis.

It will be useful for the description of these experiments if we recall some technical terms which have already been defined, namely "haploid" and "diploid" for cells or individuals which carry one or two chromosome sets, respectively, and "allelomorphs" or "alleles" for partner genes with different effects. To these terms I want to add two new ones which describe the genetical constitution of a diploid individual in regard to any particular pair of genes.

Evidently there exist two possibilities: either the partner genes are alike or they are allelomorphic. In the first case, the individual is "homozygous" or a "homozygote" in respect to the gene concerned; in the second case, it is called "heterozygous" or a "heterozygote".

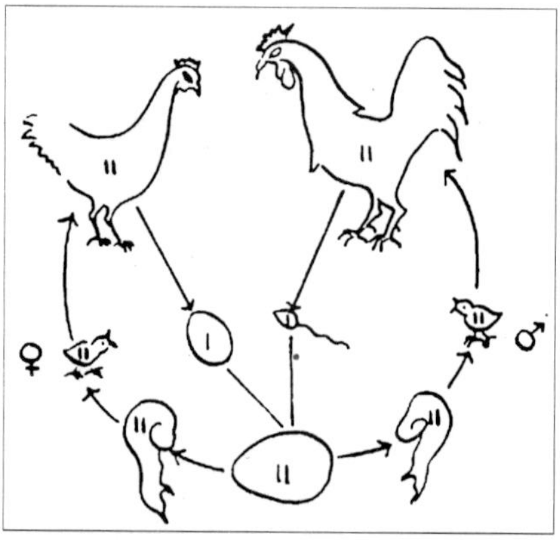

Figure: *Life Cycle of Fowl*

A pea plant with two genes for wrinkled seeds or two genes for round seeds is homozygous for either wrinkled or round; a pea plant with one gene for wrinkled seed and an allelomorphic gene for round seed is heterozygous for wrinkled and round. It will readily be seen that only heterozygotes are suitable for demonstrating the segregation of genes; for the geneticist recognizes genes by their effects, and the segregation of like genes in a homozygote is not observable.

In higher organisms (macroorganisms) it is hardly ever possible to observe the immediate effects of segregation at meiosis; for the mature gametes fuse at fertilization and the progeny, in which we observe the effects of the genes, carries genes from two different gametes. The life cycle of the domestic fowl in terms of haploidy and diploidy. The diploid female and male form haploid ova and spermatozoa. At fertilization the gametes fuse and form the fertilized ovum, which in turn develops into the diploid embryo, chicken and adult. Thus the whole life cycle, with the exception of one cell generation, is spent in the diploid stage. The situation is quite different for many microorganisms.

The Life Cycle of a Green Alga, Chlamydomonas

Each individual of this species is a microscopically small cell which swims through the water with the aid of two whiplike hairs, or "flagellae". These cells are haploid; they have only one set of chromosomes, carrying all genes of the species in single copies. New cells are formed through division of the old ones.

In this way whole populations may be derived from one cell by repeated divisions. Such populations are called "clones". Since at mitosis both daughter cells receive exactly the same set of chromosomes and genes all cells within a clone are genetically identical.

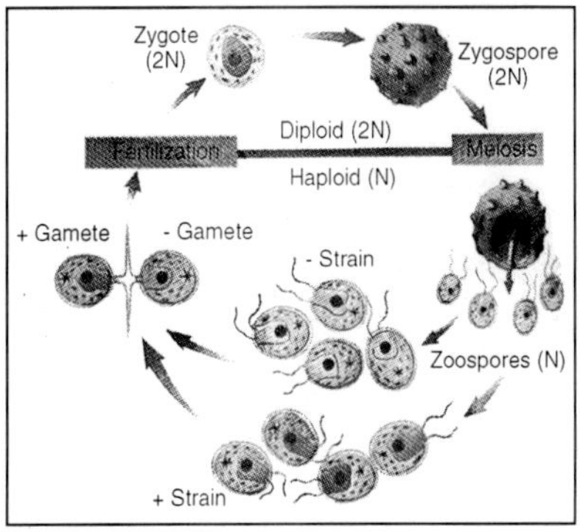

***Figure:** Life Cycle Algae Chlamydomonas*

There are no visibly distinct sexes, but physiologically there is a clear distinction between sexes or "mating types," called + and– in our diagram. Cells of the same clone are always of the same mating type and never mate with each other; nor does mating occur when cells of two + clones or two–clones are brought together. But when cells from *a*+ and *a*– strain are put into the same culture vessel, pairs are fomed.

Each pair consists of one + and one–cell they fuse and form a thick skinned pore called a zygote. The zygote is comparable to a fertilized ovum in, say, a dog or a fly. It contains the chromosomes from both gametes and is, therefore, diploid. Very soon after fertilization meiosis takes place inside the zygote, and haploid gametes swarm out.

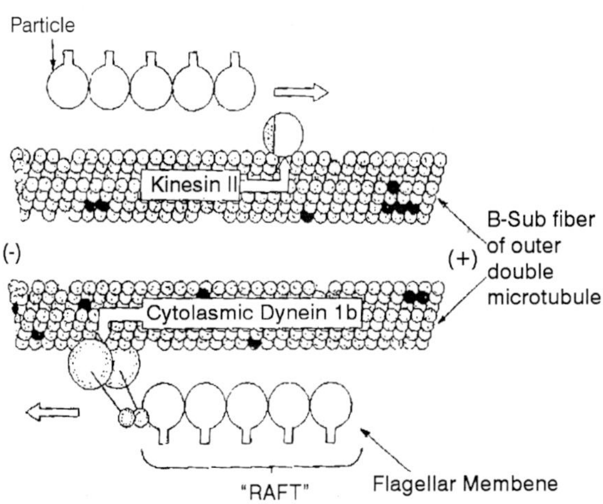

***Figure:** Absence of Flagellae in Chlamydomonas*

These start a new life cycle by division; in fact, the gametes of Chlamydomonas do not differ from any of the other free swimming cells. Thus in this algae, as in many other microorganisms, the whole life cycle, with the exception of the zygote, takes place in the haploid stage, and segregation at meiosis is not obscured by diploidy of the progeny. Usually it is necessary to grow the gametes into clones before segregation becomes observable; but occasionally the segregating gametes themselves differ visibly from each other. Two strains (1) and (2) differ in one gene which is concerned with the development of the flagellae. While strain (I) has all the genes necessary for flagella formation, strain (2) carries a deficient allele of one of these genes and cannot, therefore, develop flagellae.

When the two strains are crossed by bringing together + cells from the one strain with cells from the other (it makes no difference which strain provides the + cells, which the cells), a zygote is formed which carries both alleles and is, thus, heterozygous for the gene in question. At the reduction division the partner chromosomes segregate and carry the different alleles into separate cells. At the second meiotic division each cell divides once more by mitosis, so that finally four gametes are formed, two of which carry the normal and two the deficient allele. When these swarm out, it will be seen immediately that two have flagellae, two have not. This difference persists in the clones formed from the gametes.

If we let each gamete grow into a separate clone, two clones will consist of flagellated free moving cells, while two will consist of flagellaless cells which sink to the bottom of the tube.

The fact that a heterozygous zygote always produces two gametes carrying one allele and two carrying the other is so well established that it can be used, conversely, for determining whether a difference between two strains of a microorganism is due to the effects of a pair of allelomorphic genes.

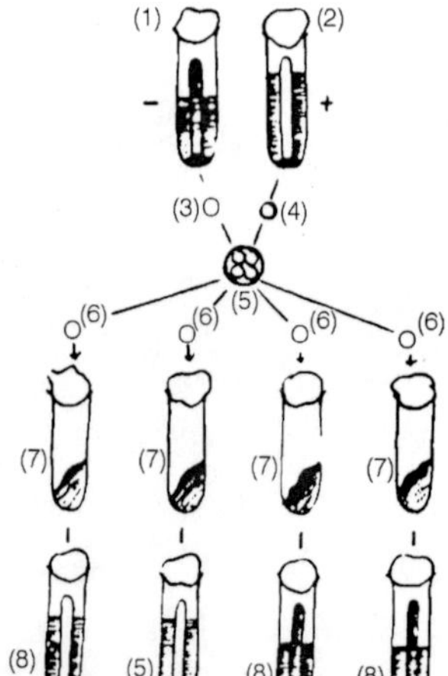

Figure: *Segregation in Yeast for the Ability to Ferment Galactose*

If this is so, the gametes from a cross between the strains will yield two clones, resembling the one strain, and two resembling the other. Yeast, like

Chlamydomonas, is a unicellular organism. Many strains are haploid and can be used in experiments on segregation. Like Chlamydomonas, yeast occurs in two mating types, + and –, which are not visibly different from each other.

Yeast, as you know, ferments sugar into alcohol and carbon dioxide gas; but strains differ in their ability to ferment various kinds of sugar. Thus some strains readily ferment galactose (one of the components of milk sugar), while others do so only very slowly and inefficiently. A simple method for detecting fermentation is to catch the developing gas in a test tube which was first completely filled with liquid and then inserted immediately above the yeast cells. If the medium contains a sugar which can be fermented, the tube gradually fills with gas; if the sugar cannot be fermented, the tube remains filled with liquid.

In the experiment shown in the medium in the top row contains galactose as the only sugar. Strain (2) is able to ferment this sugar, strain (1) is not. Cells (3) and (4) from the two strains are brought together, and a zygote (5) is formed which, through meiosis, gives rise to four gametes (6). Each gamete is grown into a clone (7) on solid medium containing glucose, a sugar which can be fermented by all strains of yeast. Finally, the four clones are tested (8) in fermentation tubes with galactose as the only sugar. You will notice that two clones (those on the left) can ferment galactose, and two cannot.

Thus the difference between the two original parents segregates in the 2:2 ratio expected if it is due to the action of a single gene pair. On the basis of several experiments of this kind it could be concluded that in yeast the ability to ferment galactose is controlled by the action of a gene which in nonfermenting strains is absent, or is present in the form of a nonfunctioning or poorly functioning allelomorph.

The essence of what is called Mendel's first law or the law of segregation. In modern terms it may be expressed as follows: When a heterozygote forms gametes, half of them carry one of the two alleles, half the other. The basis for the segregation of the genes is the segregation of the chromosomes at meiosis. Mendel himself could not, of course, interpret his results in these terms, but he quite clearly postulated that a mechanism of this kind was responsible for his data, in particular for the famous and much overrated 3:1 ratio. This particular ratio is a rather remote consequence of the underlying gametic segregation in a ratio of 1:1. The steps by which, in crosses between higher organisms, this 1:1 ratio becomes transformed into oters, including.

In doing this we shall find ourselves using very simple statistical rules, such as are used for predicting that a coin tossed 100 times will show head about 50 times. This statistical aspect of genetics has been held against it by the Lysenko school, on the grounds that mathematics must not be applied to biological objects and that Mendel had no right to speak of a 3:1 ratio between round and wrinkled peas if some of the pods contained 6 round peas. No matter what we think about these arguments—and to a modern biologist they seem entirely futile.

Characteristics of Genetic System

X-Chromosome as Type

The chromosome combinations present in the fertilized egg, and this has brought to light some of the methods by which the chromosomes act on development and characteristics. We now examine other effects of altering chromosomes. We shall have to examine the effects of altering all the different classes of chromosomesýÿthe X-chromosome, the Y-chromosome, and the different autosomes. Each of these three groups (X, Y, and autosomes) gives rise, as we shall see, to a different type of inheritance. The simplest relations, from the experimental point of view, are presented by the X-chromosomes. We therefore deal first with these.

Effects on Altering the X-Chromosomes

The X-chromosomes take such a course in passing from generation to generation that it is possible to follow the descendants of a particular Xchromosome (that present in the original male parent, for example), knowing in which individuals they are present, in which they are absent. In certain individuals (males, in organisms of Group I) there is but a single X-chromosome instead of a pair.

These relations make it a relatively simple matter to discover the distinctive effects of a particular X-chromosome. The effects of X-chromosomes will therefore be dealt with somewhat fully as a type of chromosomal action. Later the action of the other chromosomes will be taken up.

Abnormal Distribution of X-Chromosomes

What will happen if the X-chromosomes by accident become irregularly distributed? We know that sometimes the X-chromosomes are indeed irregularly distributed. Will the sex-linked characteristics continue to follow them; will the defective characteristics show the same irregular distribution as do the X's? Such cases in great number have been fully observed. It is found that the sex-linked characteristics do indeed follow the X-chromosomes wherever they go. This proves conclusively (if there were any possible doubt in view of the extraordinary course normally followed by such characters) that it is indeed the X-chromosomes on which the characters depend. The matter is one of importance and interest, so that it will be worthwhile to examine carefully certain typical cases of the result of irregular distribution of X-chromosomes.

Normal Distributions

Sometimes the body of Drosophila is yellow instead of the normal grey; this is a recessive character due to a defect in the X-chromosomes. Suppose that we mate together a female that has a yellow body and a male that has the normal grey programmes. The female has two modified X-chromosomes, which we may represent by XX. The male has its X unmodified; we may represent its cells as XY. Normally in forming the germ cells, the two X's of the female separate into different germ cells, and all the ova receive one of the modified X's. The normal male produces germ cells of two classes, X and Y. When the germ cell X from the female unites with Y from the male, sons are produced, with the constitution XY; while X from the female with X from the father gives daughters XX. Since the yellow programmes dependent on X is recessive, the daughters XX have the normal grey programmes, like the father.

But the sons XY have only one X, and as this is the defective one, from the mother, the sons are yellow, like the mother. The normal result of such a mating therefore is that all the sons are recessive like the mother, all the daughters dominant, normal, like the father. This is what occurs when the X-chromosomes are distributed to the germ cells in the usual way.

Abnormal Distribution

In forming germ cells both X's of the female go together to a single ovum, while other ova receive no X. In a race discovered by L. V. Morgan the two X-chromosomes in the female were partly united, so that they almost invariably thus go together to one ovum, leaving other ova without an X. In this race the X-chromosomes were so modified as to produce the recessive yellow programmes described in the preceding paragraph. We therefore have an opportunity to determine the course of inheritance of the yellow and the grey programmes when the X-chromosomes are thus abnormally distributed.

Represent as before the X-chromosomes that produce the yellow programmes by X. Thus one set of ova receive XX, while another set receive no X. When the germ cells XX united with the germ cells Y from the normal father, their wee produced daughters XXY, in which both X's came from the mother, and these daughters were yellow like the mother (since no normal X was present). Sons were produced by the union of ova that contained no X with sperms that carried X; this gave sons XO, with their single X from the normal father, instead of from the mother, as is usually the case. And such sons had the normal grey body programmes of the father, instead of the yellow body programmes of the mother, as happens normally. Thus with a change in the distribtion of the X's there is a corresponding change in the method of inheritance. If the sons receive their X from the recesive mother, as in the normal cases, they are recessive lke the mother. But when, through non-dijunction, the sons receive a dominant X from the father, they are dominant like the father.

Similarly, in the normal case the daughters receive a dominant X from the father and are therefore dominant like the father. But in cases of non-disjunction the daughters receive both their JPs from the recessive mother, and are therefore recessive like the mother. In sum, when the X's are normally distributed, such matings give 'criss-cross inheritance'; sons like the mother, daughters like the father. But when the X's are not so distributed, there is no criss-cross inheritance; sons are like the father, daughters like the mother. The dominant and recessive characters follow the respective X-chromosomes, whether these are distributed normally or abnormally.

Such experiments have been repeated many times, and with other sex-linked characters. Always the characteristics follow the distribution of the X-chromosomes, however these are distributed. It may be concuded with certainty that it is the presence of X-chromosomes of a modified type that causes the appearance of the particular sex-linked characters that remanifested. The rules of distribution of sex-linked characters are the rules of distribution of the X-chromosomes.

Dominance and Recessiveness in Sex-linked Characters

As we have seen, and shall see further, many defective conditions are due to defects in certain X-chromosomes. These follow from generation to generation the distribution of these X-chromosomes. Most of these bodily defects are manifested only in individuals in which the defective X- chromosomes are the only kind present; that is, these defects are recessive.

But it is important to observe that in such cases the normal condition of the organism likewise follows the distribution of certain X-chromosomes. Haemophilia follows certain X- chromosomes. But in the same matings in which this occurs, the healthy condition of the blood follows certain other X-chromosomes. If a defective mother is mated with a normal father, the sons are defective because they receive only the mother's defective X-chromosomes.

But in the same way the daughters are normal because they receive the father's normal X-chromosomes. In the normal individuals the X-chromosome plays a part, just as it does in the defective individuals. In the normal individuals it is healthy and supplies what is required for normal development, while in the defective individuas it fails to supply what is required. In the case of most defects this normal condition is dominant over the recessive condition; that is, when both are present, the normal chromosome is the one that prevails in its effect on the indvidual. There are defective characters, however, in which the defective condition is dominant or partly dominant.

In these cases, when a defective X and a normal one are present together, the defective one produces its effect, as in the case of bar-eye, described earlier. In many such cases the defect is less marked when a normal chromosome is present as well as a defective one; then the defective condition is said to be partly dominant. Thus whenever an individual bearing defective X-chromosomes is mated with a normal individual, we have both dominant and recessive characters, the course of which may be followed in the later generations the dominant character being commonly the normal or usual condition. How this works out is shown in our next paragraphs.

Tests for Sex-linked Inheritance

By observing the distribution of the dominant and recessive characters among the offspring of certain matings, a test for determining whether given characters depend on the X-chromosomes is supplied. The test consists in making what are called reciprocal crosses.

- On the one hand mate a dominant female with a recessive male;
- Also mate a recessive female with a dominant male.

Represent a dominant X-chromosome by a capital X, a recessive X-chromosome by a lower-case letter x. In the group of organisms with which we have been dealing, the female has two X-chromosomes, the male but one. The daughters receive an X from each parent, the sons an X from the mother only. A dominant female will be represented by XX, a recessive male by xo; similarly a recessive female is xx, a dominant male is XO.

The two matings and their results will then be represented as follows:

- XX by xo gives X + XO Dom Mother Rec. Father Dom. Daughters Dom. Sons
- xx by XO gives X + xo Rec. Mother Dom. Father Dom. Daughters Rec. Sons

Thus when the mother is dominant, the father recessive, all the children are dominant, like the mother. When the father is dominant, the mother recessive, the daughters are dominant like the father, the sons recessive like the mother; this is called 'criss-cross inheritance'. Whenever reciprocal crosses give these results, it is certain that the two diverse characteristics (dominant and recessive) result from dfferences in the X-chromosomes of the two parents. 'Criss-cross inheritance' is particularly useful in showing at once that we are dealing with sex-linked inheritace.

In every case where such results are produced, further tests show that the two haracteristics follow in later generations the two different kinds of X-chromosomes wherever they go. It is mainly by the use of these tests that the many different characteristics dependent on diversities in X-chromosomes have been discovered.

Sex-linked Inheritance in Group II

Sex-linked inheritance as we have just described it was originally discovered in animals belonging to Group I, in which the females have two X-chromosomes, the males but one (with or without a Y- cromosome). But when the tests we have just described were applied to birds and to certain other animals, it was discovered that there is a second group (Group II) in which it is the male that has two X-chromosomes, the female but one. In th common fowl, reciprocal crosses were made between 'barred' fowls (Plymouth Rock) and black fowls (Langshan). Here the 'barred condition, was found to be dominant.

The results were:

- Barred father by black mother=Barred sons + barred daughters.
- Black father by barred mother = Barred sons + black daughters.

Here the second mating gives 'criss-cross inheritance', showing that we are dealing with sex-linked characters dependent on differences between the X-chromosomes of the parents. But, in the first mating above, all the offspring are like the father instead of like the mother, which latter was the case in the matings. The father in this case has the dominant characteristic (barred), and all the offspring are dominant like the father instead of like the mother.

When the mother is dominant ('barred') the result in the fowls is criss-cross inheritance while, in the cases before described, criss-cross inheritance occurs when it is the father that is domiant. In fact, in the birds, the males and females simply exchange roles, as compared with their roles in the organisms of Group I. Since the roles of males and females are interchanged in the two groups, their chromosomal conditions must also be interchanged; that is, since in Group I the female has two X's, and in Group II the male plays the same role as the female of Group I, the male of Group II must have two X's, the female but one. The chromosomal conditions in the two group are shown in the above tabulation, X signifying dominant, while x signifies recessive, and o signifies the lack of X (whether Y is present or not).

Careful examination shows that this is the only way in which the results in Group II can be produced. Birds and certain mothers show inheritance of the kind typical for Group II. In both these it has since been found under the microscope that the males have indeed one more chromosome than the females.

Role of Development in X-Chromosomes

A great many characteristics, of various kinds, have been found to be the result of modifications of particular X-chromosomes, and thus to follow in later generations the distribution of the descendants of those chromosomes. Examination of a number of these is desirable, both for their own importance and for the light they throw on the functions of the X-chromosomes.

In man, the following characteristics, among others, dependent on diverse types of X-chromosomes, are known from the fact that they show typical sex-linked inheritance.

- *Haemophilia*: Lack of coagulability in the blood. This results from a serious defect in certain X-chromosomes. The existence of defective X-chromosomes having this result shows that the normal X-chromosomes play a part in supplying something necessay for producing normal blood that coagulates properly.
- *Programmes Blindness*: The fact that defectiveness in X-chromosomes causes programmes blindness shows that the normal X's play a role in producing the normally functioning eyes.

Night Blindness

- Inability to see in a poor light.
- Near-Sightedness of certain types.

Progressive atrophy of the muscles: The fact that defects in X-chromosomes have this effect shows that the normal X's play a role in the normal functioning of the nerves and muscles. A considerable number of other sex-linked characters are known in man. But man is a very unfavourable organism for the study of inheritace. Yet even the little that is known of sex-linked inheritance in man shows that the X-chromosome plays a role in many diverse bodily functions.

For a more complete idea of its role, some organism that can be bred experimentally must be examined. For this urpose the fruit-fly, Drosophila melanogaster, is the best organism to select, since it has been studied more extensively than any other. In the fruit-fly we find peculiarities of the following types that are dependent on alterations in particular X-chromosomes:

- *Many different eye colours*: The normal eye programmes in this organism is a certain shade of red. Defects in different X-chromosomes result in producing, in the individuals that bear these defective chromosomes, many different shades of red, varying from deep red to a very light red, and thence to 'buff', 'ivory, and 'white'. More than a dozen different types of eye programmes are known to result from modifications of the X-chromosome. It is clear that the normal X-chromosome plays an important role in producing the normal eye programmes. Structural peculiarities of the eye, such as 'bar-eye', 'facet eye', 'furrowed eye', and the like, are known to result from modifications of the normal X-chromosome.

- *Wing modifications*: Many different conditions of the wings are known that depend upon diversities among X- chromosomes borne by different individuals. These affect all sorts of features of the wings: size, form, venation, function. The normal X-chromosome obviously plays an important role in the full and normal development of the wings.
- *Body colours and markings*: Modifications of the X- chromosomes produce the body colours yellow, sable, tan, chrome, lemon, green, and the like, in place of the normal grey. Other changes in X alter the distribution of pigment on the body, giving the characters dot.
- *Body structure*: A defect in X produces irregularities in the abdomen, known as 'abnormal abdomen'.
- *Legs*: A defect in X-chromosomes causes abnormal development of the legs; some of them are wholly or partly reduplicated, so that the total number of legs is increased.
- *Bristles*: Modifications in X-chromosomes result in various different changes in the bristles that are scatered over the body.

Many physiological conditions and functions likewise depend upon the X-chromosomes, since they are change when these chromosomes are modified. Among these are the following:

- *Positive reaction to light*: rosophila with normal X-chromosomes fly towards a source of light. Those having defective X-chromosomes of the kind that produce a tan-coloured body do not fly towards a source of light.
- *Weakness and short life*: Most of the defective X-chromosmes that cause structural or other changes in the body (different eye colours, wing forms, and the like) produce likewise weakness and short life. The individuals bearing them are less resistant to bad conditions, and live a shorter time than the individuals that bear the normal X-chromosomes.
- *Life and death*: Some X-chromosomes have defects that are so severe that the individuals bearing them will not live and develop, unless there is resent also a normal X-chromosome. Such defects are known as lethals. The presence of such lethal defects in the X-chromosome has the result that males bearing them die, since they have but one X; while females usually live, since they have an additional X that is often normal.

If a mother that has one lethal X and one normal one is mated to a father that has a normal X, the result is that half of the sons which receive the mother's defective X fail to develop, while the other half of the sons,receiving the mother's normal X, live and develop. All the daughters live, since they all receive a normal X-chromosome from the father. The consequence is that in such families there are twice as many daughters as there are sons. In families of 100 to 200, as occur in Drosophila, this is very striking. X-chromosomes are known to play in other organisms roles similar to those mentioned above for Drosophila, although in no other organism has the matter been so fully studied.

We may summarize what has been brought out above in the following statemets:

- Many defects that appear in individuals are the result of defects in the X-chromosomes that they bear.
- Some of these defects redominant; they are manifested in all individuals in which the defective X is present.
- But most such defects are recessive; they are manifested only in individuals in which the defective X is the only kind present.
- When both a normal X and a defective X are present, in most cases the normal X performs the required functions, so that the individual is not defective.
- Since the female has two X-chromosomes, while the male has but one, such defects are more frequently manifested in the males. Usually one of the X's present in the female is without the defect, so that she is not defective. But when the male carries a defective X-chromosome, the defect is manifested.
- It is thus advantgeous, so far as the occurrence of defects is concerned, to have two X-chromosomes rather than one.
- This gives the female a considerable advantage over the male in these respects. There are many different types of defects due to defective X-chromosomes; almost all of them are more common in males than in females. Some of these defects seriously injure the health, or even result in death, if the defective X is the only kind that is present. It is well known that, in general, the death rate is higher in males than in females; this is probably due to the fact that males have but one X-chromosome.
- Among the different individuals of a species are scattered a great number of different types of X-chromosomes having different effects on development. Many of the diverse X-chromosomes are distinctly defective, causing personal defects in the individuals that bear them. Others are not defective.
- Defects or modifications in X-chromosómes affect in different cases all parts and functions of the organism.

The Role of the X-Chromosomes in Development

What do the facts brought out in the preceding sections show as to the phyiological functions of the X-chromosomes; as to their role in development? For every kind of effect resulting from defects or modificacations in the X-chromosomes, there is a corresponding (opposed or diverse)action of normal or unmodified X-chromosomes. Since defective X-chromosomes of a certain type produce in Drosophila short, ill-developed wings, it follows that the normal X-chromosomes act on the development in such a way as to give long, well-developed wings.For if we substitute the normal X for the defective one, this cause the normal well-developed wings to be produced.

Similar reasoning applies to all the defective conditions that result from defective X's; the normal X's act in such a way as to produce the corresponding normal conditions. It follows therefore that the normal X-chromosomes play a role in the production of eye programmes and structure, body programmes and structure, in the development of the legs, the wings, the bristles, in producing normal health, resistance and vigour, and in various physiological processes. We know, further, that they play a most important part in determining sex, with all that this includes of structural and physiological influence. Clearly therefore the X-chromosomes play a most important part in development. Their action is not limited to any single part of the body, nor to any single class of functions. They enter into the processes of development in such a way as to influence all parts of the body, all functions of the body. They begin to influence development very early, as was seen in considering the development of sex. And they play important roles in such relatively late processes as the production of eye colours. It can hardly be doubted that they enter into the developmental activities from practically the beginning, influencing all the later processes that occur.

Autosomes and T-Chromosomes

The X-chromosomes have been dealt with because of the great advantages they offer for experimental study. It was seen that they affect development in many ways, and that in a given species there are many different types of X-chromosomes in the different individuals, causing them to show different characteristics, structural and physiological. Is the X-chromosoe typical in these relations? Shall we find similar effects in the other chromosomes? In addition to the X-chromosomes, there are the autosomes, which are present as a rule as several or many pairs. In man there are 23 pairs of autosomes: in Drosophila there are three pairs. There is, further, in many species the single Y-chromosome. The autosomes and the Y-chrmosome are distributed from parent to offspring in characteristic ways, differing from the distribution of X-chromosomes. If they affect characteristics, these differences should appear in a different system of inheritance. We shall examine first the autosomes.

The Autosomes: Typical Mendelian Inheritance

We have already seen that the autosomes play a role in development, in the fact that they help to determine the sex of the developing individual. In Drosophila, if the number of sets of autosomes, in relation to the number of X's, is changed, this changes the sex. Do autosomes also affect other characteristics? Do the diferent pairs of autosomes have different-functions? Are there diverse types of autosomes, with different effects, in different individuals of a species, as is the case in relation to the X-chromosomes? Are there dominant and recessie effects of autosomes, as there are of X? If dominant and recessive characteristics depend on auosomes, then these characteristics must follow the distribution of the autosomes from parents to offspring, as the characteristics dependent on X follow its distribution. We must then examine the method of distribution of the autosomes whether there are characteristics that show this method.

Method of Distribution of the Autosomes

The ditribution of the autosomes from parents to offspring is much simpler than that of the X-chromosomes.

The main facts are as follows:

- The autsomes are in pairs in all individuals. There are no individuals with unpaired autosomes, like the male of some organisms with relation to X.
- Thus the two sexes are alike as to their autosomes.
- In forming germ cells, one member of each pair of autosomes goes to each germ cell. The sperms and the ova are alike with respect to the autosomes.

  Thus the fundamental rule with relation to the autosomes is that each gem cell carries one member of each pair of autosomes.
- By union of two germ cells the pairs of autosomes are restored.
- Thus each one of the offspring carries in each pair of its autosomes one autosome from each parent.

Now consider a single pair of autosomes, such as the pair marked AA in. Suppose that in different individuals the members of this pair of autosomes produce different effects, as we have seen to be the case with the X-chromosomes of different individuals. And suppose that, as in the case of X, one of these effects is dominant, the other recessive. How would such characteristics be inherited? This question must be examined with care. For this purpose we may take as an example the different colours found in different varieties of peas, used in the classic experiments of Mendel. The yellow seed programmes of a certain variety is dominant, and results from the constitution of the autosomes of a certain pair. The green seed programmes of another variety is recessive, owing to a different constitution of the autosomes of that same pair.

Call the two autosomes of that pair in the dominant yellow race by the capital letters *AA*, while the two autosomes that give rise to the recessive green programmes in the other race may be called *aa*. The dominant parent may be designated *D*, the receive parent *R*, When these two parents produce germ cells, each germ cell receives, of course, one chromosome for the pair. The germ cells from the dominant parent *D* have the autosome *A*, those from the recessive parent *R* have the autosome *a*. The germ cells *A*, from parent *D*, unite each with a germ cell a from parent *R* giving offspring (zygotes) containing both *A* and *a*. These first generation offspring are commonly designated *Fi* (first filial generation). And since the characteristic carried by the autosome *A* is dominant, all these *Fi* offspring manifest the dominant characteristic *D* (yellow, in case yellow-seeded peas are crossed with green-seeded peas). Next observe the result of mating together two of these individuals *Aa*, of the *Fi* generation, giving the *Fa* (second filial generation).

Each parent *Aa* gives, according to the general rule, two kinds of germ cells, one kind carrying the dominant autosome *A*, the other the recessive autosome *a*. Each kind of germ cell from one parent unites with each kind from the other. That is, half the *A* germ cells from one parent unite with *A* germ cells from the other parent, half with a germ cells from the other parent; and the same is true for the *a* germ cells from the first parent.

This gives in equal proportions Fa offspring (zygotes) as follows:

$$AA + Aa + a\,A + aa$$

But as the two combinations *Aa* and *aA* contain the same chromosomes, the proportions may be written *AA* + 2*Aa-f-aa*. The individuals of the constitution AA and Aa contain the dominant chromosome *A*; therefore they manifest the dominant characteristic, *D*. The individuals *aa*, having only the recessive chromosome *a*, manifest the recessive characteristic, *R*. Thus in this F^2 generation, so far as manifested characters are concerned, the proportion of offspring are 3 Dominant to *i* Recessive

Thus if dominant and recessive characteristics depend on the two members of a pair of autosomes, then a dominant individual *AA* is mated with a recessive individual *aa*. The general result is bound to be that the immediate offspring (*Fi*) all manifest the dominant characteristic, while in the *Fa* generation (produced by mating two individuals from *Fi*) there will be three times as many dominant individuals as there are recessives. Now, this is the result that was discovered by Mendel in 1866; it formed the foundation for the working out of Mendelian heredity. The proportion 30 to *iR* in the *Fa* generation is commonly known as 'the Mendelian ratio'; it is one of the most characteristic proportions in this type of inheritance.

Mendel discovered that there are many different characteristics that yield this 3 to *i* ratio. If yellow and green peas are crossed, in the *Fa* generation there are 3 yellows to *i* green; if round peas and wrinkled peas are crossed, in F^2 there are 3 round to *i* wrinkled; and so for may other characteristics in the pea plants with which Mendel worked.

And since the time of Mendel these same ratios have been found for thousands of different characteristics, in great numbers of different organisms, from the pea plant and *D* rosophila to man. The fact that inheritance gives these ratios is commonly spoken of as Mendel's law, and the type of inheritance which it exemplifies is spoken of as Mendelian inheritance. The type that gives in the *Fa* generation 3*D* to *iR* is the commonest method of inheritance, much commoner than the sex-linked method, in which the characteristics depend on differences in the X-chromosomes. This is obviously because autosomes are more numerous than X-chromosomes.

We find then that inheritance according to Mendel's law takes place exactly as it would if the characteristics concerned depended on the two members of a pair of autosomes. As will be shown later, it can be proved experimentally that such characteristics do indeed depend upon autosomes. Thus the rules of Mendelism are the rules of distribution of the two members of a pair of autosomes, one having a recessive effect, the other a dominant effect. Mendelian heredity is heredity of characteristics that depend on differences in autosomes.

Mendel did not know of the relation of the characteristics to the chromosomes. He discovered, however, certain other important facts about this type of heredity, facts which have become clearly explicable since it has been learned that this type depends on differences in autosomes. These facts must be examined. For this purpose it will be convenient first to define certain terms which are universally employed in dealing with inheritance.

By the union of two germ cells or gametes, as *A* and *a*, there is produced a zygote. If the two germ cells are alike with respect to the chromosomes or characteristics that they carry, the zygote so produced is spoken of as a homozygote; thus *AA* and *aa* are homozygotes. If the two germ cells that unite to form a zygote are unlike, as *A* and *a*, the zygote so produced is called a heterozygote, as *Aa*. The two corresponding characters, one dominant, the other recessive, that are produced by the two differing members of a pair of chromosomes are spoken of a alternative characters, or alleles, or allelomorphs. Any given individual manïfests one or the other of the two alternatives, not both. For the alternative characters we shall employ henceforth the term alleles.

The characteristic manifested by a zygote is its phaenotype while its actual constitution is its genotype. Thus, two zygotes *AA* and *Aa* both manifest the dominant character; the dominant character is the phaenotype, or the individuals are said to be phaenotypically dominant. But they are diverse in genotype; one is *AA*, the other *Aa*. In the case of a homo-zygote *AA* or *aa*, the phaenotype and genotype may be held to coincide. Mendel crossed yellow peas with green peas, and found, as before stated, that in the *Fi* generation all the peas were yellow, so that yellow is the dominant allelomorph. In the F^2 there were three times as many yellows as greens. The actual number of seeds in the F^2 generation was in his experiments 8023, of which 6022 were yellow while 2001 were green, so that the ratio of dominants to recessives was, accurately expressed, 3.01 to *i*.

Mendel found that if he bred further the individuals of the F^2 generation, some of these gave different results from others. In the pea plant it is possible to allow each plant to fertilize itself, so that both kinds of germ cells come from the same parent. When this was done with the individuals of the F^2 generation, the following results were observed: The green plants (recessive) gave only recessive green offspring (they 'bred true'). Of the yellow plants (dominant), one-third gave only yellow offspring (they 'bred true'), while two-thirds of them gave both yellow and green offspring, in the proportion of 3 yellows to *i* green.

The reason for this is clear when we consider the chromosomes on which the characters depend. If *A* represents the chromosome (autosome) that gives the dominant yellow programmes, while a represents the chromosome that gives the recessive green, the original parents were *AA* and *aa*, the *Fi* generation were all *Aa*, while the *Fa* generation showed the roportions *AA-f* ^{2} *Aa* + *aa*.

The recessive individuals *aa* are homozygotes; their two chromosomes are alike, being *aa*. When germ cells are produced by these homozygotic individuals, all these germ cells must contain the chromosome *a*, and when two of these unite in self-fertilization, of course they produce in all cases again individuals *aa*. These are, of course, recessives (green), like their parents; that is, the green parents aa 'breed true'.

Similarly the dominant yellows, *AA* (constituting one-third of all the dominants), are homozygotes. Their germ cells all contain the chromosome *A*, and the offspring produced by union of such germ cells, in self-fertilization, are all *AA*, and so doinant yellow like the parents. One-third of all the dominants thus breed true. But the individuals *Aa* are heterozygotic; that is, they have two kinds of chromosomes, *A* and *a*, in the pair.

Thus they produce two kinds of germ cells, *A* and *a*, in equal numbers. And when these germ cells unite, they necessarily give, as we saw on an earlier page, offspring of three kinds, in the proportions *AA* + 2*Aa*4-*aa* Of these, *AA* and 2*Aa* will be dominants (yellow), while *aa* will be recessive (green). Thus the homozygotes *aa* and *AA* breed true, while the heterozygotes, *Aa*, do not.

As Mendel discovered, if the heterozygotes *Aa* are mated with the recessive homozygotes *aa*, they give two kinds of offspring, *Aa* and *aa*; half (*Aa*) being dominant,the other half (*aa*) being recessive. If the heterozygotes *Aa* are mated with the dominant homozygotes *AA*, the offspring *AA* and *Aa* are all dominant, though half (*AA*) are homozyotes, the other half (*Aa*) heterozygotes. All these results are typical for the descendants of two parents different in a single pair of chromosomes, one pair being dominant, the other recessive. For reference it will be well to make a table of these results, characteristic for Mendelian inheritance.

The Results of Mendelian Inheritance

Parents: *AA* and *aa*; both homozygotic. *Fi*: *Aa*; all alike, dominant heterozygotes. *Fs*: *AA* + 2*Aa* + *aa*. Three-fourths of the individuals dominant, one-fourth recessive. Half the individuals homozygotes (*AA* and *aa*), the other half heterozygotes (*aAa*). Each type of homozygote bred by itself produces offspring like itself. The heterozygotes *Aa*, when bred by themselves, produce again offspring. *AA* + 2*Aa* + *aa Aa* mated with αα yields again *Aa* + *aa Aa* mated with *AA* yields again *Aa* + *AA*. All of these results are found to hold true for thousands of different characteristics.

We may therefore summarize the situation in the following general propositions:

- When two different individuals are crossed, in many cases the differences between them follow the same method of distribution as do the two members of a pair of autosomes ýÿone being dominant, the other recessive.
- This distribution is what is known as typical Mendelian heredity.
- The two sexes do not differ in these respects. Each sex has the same proportion of dominants and recessives.
- The rules of Mendelian inheritance are essentially the rules of distribution of the members of pairs of autosomes.

Characteristics that depend on two or more pairs of autosomes: We have thus far followed the distribution of characteristics that depend on differences in a single pair of autosomes. But most organisms have several or many pairs of autosomes: the fruit-fly has three pairs, man has twenty-three.

Do some characteristics depend on differences in one pair of autosomes, others on differences in other pairs? Have the different pairs of autosomes thus different functions in development and inheritance? We find that the different pairs of autosomes have indeed different functions, and that some characteristics depend on one pair, some on another. This is shown in many cases in which two parents differ in two diverse characteristics. One characteristic may then act as if dependent on one pair of autosomes, the other on another pair. As an example of this, we may take a case studied by Mendel in peas. One of the original pea plants had seeds that were (1) yellow, and (2) round, or smooth. These, as it turns out, are both dominant characters. The other plants, crossed with those just mentioned, had recessive characters; the seeds were (1) green, and (2) wrinkled.

When these two kinds of pea plants were mated together, it turned out, as we shall see, that yellow and green behave in inheritances as if due to a difference in the two parents in the effects of a certain pair of autosomes; that is, they are alleles. Round and wrinkled be have as if due to a difference between the two parents in the effects of another pair of autosomes. We may therefore represent the two pairs of autosomes. Representing the dominant characters by capitals, the recessives by lower-case letters, we may in the diagram represent the chromosomes giving origin to the four characters as follows:

A = yellow *a* = green

B = round *b* = wrinkled

The germ cells from these parents contain, of course, one member of each pair. One parent produces gametes all of the constitution *AB*, the other gametes all of the constitution *ab*. When a gamete *AB* from one parent unites with a gamete *ab* from the other, all the offspring (*Fi*) thus produced have the constitution *AaBa*, as shown at *Fi* in Fig.. Since the dominant autosomes *A* and *B* are both present, all these individuals are domiant for both pairs of characters; that is, they all have seeds that are yellow and round, as Mendel found. But all are heterozygotes-for both of these characters.

Next we mate together two of these *Fi* individuals to give the F^2 generation. In forming gametes the rule is, of course, that each gamete contains one member of each pair of chromosomes. But the two pairs, *Aa* and *Bb*, are distributed to the gametes quite independently. Thus, some get *A* and *B*, some *A* and *b*, some α and *B*, some *a* and *b*. There are thus from each parent four types of gametes, *AB, Ab, aB*, and *ab*, each type being represented by many gametes. The gametes of each of the four types from one parent unite in approximately equal proportions with gametes of each of the four types from the other parent. That is, gametes of the type *AB* unite in different cases with *AB, Ab, aB* or *ab* from the other parent; and so of each of the other gametes of the first parent. The result is to give a set of 16 combinations of gametes (some of which are alike, however, as will be seen).

Inheritance Depending on Two Pairs of Autosomes

AB AB AB AB

AB(i) *Ab*(6) *aB*(7) *ab*(s)

Ab Ab Ab Ab

AB(6) *Ab*(2) *aB*(5) *ab*(8)

aB aB aB aB

AB(7) *Ab*(5) *aB*(3) *ab*(9)

ab ab ab ab

AB(5) *Ab*(8) *aB*(9) *ab*(4)

But, as will be observed, not all these 16 are diverse; thus *ABAb* is the same in constitution or genotype as *AbAB*. If we number the different combinations, and give the same number to those that are alike, as is done (in parentheses) we find that there are nine different combinations or genotypes. That is, two parents that are both *ABab* produce offspring of nine different constitutions or genotypes.

When the parents differed in but one pair of characters there were but 3 different combinations. When there are two pairs the number is 3 > or 9⁻.

These nine different combinations will manifest the dominant characters *A* or *B*, in all cases where an *A* or *B* is present whether present once, as *A*, or twice, as *AA*); they will manifest the recessive characters *a* or *b* in all cases in which *a* or *b* is present twice, no corresponding capital letter being present. Collecting together these different types, we find the different characteristics to be manifested in the following proportions of the zygotes:

$$9AB + 3Ab + saB + 1ab$$

That is, 9 out of 16 are dominant for both pairs of characters, 3 are dominant for *A*, recessive for *b*; 3 dominant for *B*, recessive for *a*, and *i* out of 16 is recessive for both. There are thus 4 different phaenotypes.

In Mendel's experiment with peas, in which the two pairs of characters were:

1. Yellow and green;
2. Round and wrinkled; in *Fi* the proportions were 9 that were yellow and round,
3. That were yellow and wrinkled,
4. That were green and round, and *i* that was green and wrinkled.

These proportions hold generally when there are two pairs of characters connected with different pairs of autosomes.

It may be observed that the proportions of the individuals of different constitution or genotype in F^2, descended from a cross between parents that differ in respect to two pairs of autosomal characters (so that the parents are *AABB* and *aabb*), can be found directly from the proportions for a single pair of characters. In *Fa*, for the single pairs the proportions are *AA* + 2*Aa*-*haA* and *BB* + 2*Bb* + *bb* respectively. If now we multiply together algebraically these two polynomials, we obtain the relative proportions for the different combinations produced when the two pairs are together.

These proportions are the following:

$$AABB + 2AaBB + aaBB + 2AABb + 4AaBb +$$

(1) (7) (3) (6) (5)

$$2aaBb + AAbb + 2Aabb + aabb$$

(9) (2) (8) (4),

Certain other relations among the nine different types are important, since they hold for all cases of two pairs of characters that are dependent upon different pairs of autosomes.

These are the following:

- There are four combinations (Nos. 1, 2, 3, 4), constituting one-fourth of all the zygotes, that are homozygotes for both pairs of gens. The four homozygotes are diverse in their combinations and in the characters they manifest. Any one of these when self-fertilized or mated with another like itself breeds true: that is, it produces offspring that have the same constitution as the parents.
- Four out of the 16, or one-fourth of all the zygotes, are heterozygotes for both pairs of characters. They form the class numbered (5), and constitute the diagonal from upper right to lower left. Any of these when self-fertilized or mated with another like itself gives again the nine different types, in the proportions above set forth.
- The one-half of all the zygotes, are homozygotes for one of the two pairs, heterozygotes for the other. Four are homozygotic for *A* and heterozygotic for *B*; the other four are heterzygotic for *A* and homozygotic for *B*. In these 8 there are but 4 diverse combinations or genotypes.

If any one of these is self-fertilized, it yields 3 diverse types with respect to the pair for which it is heterozygotic. Thus, the class (6), having the constitution AABb, yields when self- fertilized three combinations in the following proportions:

$$AABB + 2AABb + AAbb$$

The above relations hold for all cases in which two pairs of characteristics are dependent on differences in two different pairs of autosomes, whatever the method of designating the characters or chromosomes.

In a similar way, two parents may differ in three characters that are dependent on three different pairs of autosomes. Such parents could be represented as *AABBCC* and *aabbcc*.

Their progeny (*Fi* generation) would all have the constitution *AaBbCc*. Such *Fi* individuals of course produce a variety of germ cells, in accordance with the principles that each germ cell receives one member of each pair of autosomes, and that the different pairs of autosomes are distributed independently. Each parent *AaBbCc* thus yields the following 8 types of germ cells:

ABC, ABc, AbC, aBC, Abe, aBc, abC, abc.

When each of the 8 kinds from one parent unites with each of the 8 kinds from the other, there are produced, of course, 64 groups. If we classify the 64 individuals by the characters which they manifest, we find that there are 8 different phae-notypes manifesting respectively the following characteristics:

ABC, ABc, AbC, aBC, Abc, aBc, abC, and *abc*

That is, some are dominant for all three characters; some for a particular two and recessive for the other; some are dominant for a particular one and recessive for the other two; and some are recessive for all three. The different phaeno-types are present in different proportions; the proportions when a large number are examined are as follows:

2ABC + gABc + gAbC + gaBC + $Abc + saBc + 3abC + iabc

The total number of different combinations (genotypes) that occur in such a case is. These combinations can be obtained in their correct proportions by multiplying together algebraically the expressions: Similar but still more complex series are obtained if parents are interbred that differ in four or more characters depending on four or more different pairs of autosomes. If the number of pairs of autosomes is called ↔, the following relations hold:

If parents differ in n autosomes:

- The number of diverse types of gametes produced by the individuals of the *Fi* generation is 2^n.
- The number of different combinations (genotypes) in the zygotes of F^2, formed by mating the 2^n gametes from *Fi*, is s^w.
- The number of different phaenotypes produced in *Fa* is 2^n.

These relations yield very great numbers of diverse combinations, in cases where the number of pairs of autosomes is large. Thus in man, with 23 pairs of autosomes, if all the autosomes differed in the two parents, the number of types of germ cells produced by the *Fi* individuals would be 2, or many millions, the number of possible diverse genotypes in F^2 would be 3, or many billions, and the number of possible different phaenotypes would be 2. (As will appear later, the number of different possible combinations is greatly increased by the exchange of parts by the members of a pair of chromosomes.)

Thus typical Mendelian inheritance, dependent on differences between different pairs of autosomes, can be treated to a certain extent mathematically. The individuals in F^2 and later generations occur in proportions represented by simple mathematical ratios.

It will be worthwhile to formulate the main facts that we have brought out as to autosomal inheritance:

- Some single pairs of characters, dominant and recessive, follow in inheritance the method of distribution of a single pair of autosomes.
- Some pairs of inherited characters follow the rules of distribution of two or more different pairs of autosomes, in the following particulars:
  - — One member of each pair goes into each germ cell.
  - — The members of the different pairs are distributed to the germ cells independently, so that all possible combinations containing one member of each pair are formed with equal frequency.
  - — Each type of gamete from one parent unites with each type from he other parent, the different matings being equally frequent.
- These methods of distribution result, for the characters manifested in the Fs generation, in certain typical numerical proportions or ratios, known as Mendelian ratios.
- This method of inheritance is known as Mendelism, or Mendelian inheritance. The rules of Mendelism are the result of the rules of distribution of pairs of autosomes.
- Most characteristics of organisms are inherited in this way; many more than are inherited in the sex-linked manner. This is evidently because there is but a single pair of X-chromosomes, while there are several or many pairs of autosomes.
- Thus it is clear that, among the different individuals of a species, there are many different types of autosomes (as there are different types of X-chromosomes), producing many kinds of characteristics.

The T-Chromosome

There remains another chromosome, not included in the autosomes: namely, the Y-chromosome. Has this functions similar to those of X and of the autosomes? Does it too play a role in heredity? As we shall see, the Y-chromosome differs in many ways from the other chromosomes. Let us first summarize the main facts as to the occurrence and distribution of Y-chromosomes.

- Some organisms have no Y-chromosomes. Among those are many insects the dog, the cat, the horse.
- In some organisms it is small and appears to be degenerate. This is the case in many insects; also in man.
- In some organisms the Y-chromosome is large, but of a different form from its mate X. Such is the case in Drosophila

- In some organisms Y is not distinguishable from X inform or size, but experiment shows that it differs from X in function.
- The Y-chromosome descends from father to son only, exclusively in the male line. Normally it never occurs in females.
- Hence any characters that depend on the presence of a particular kind of Y will *(a)* be found in males only, and will *(b)* never be inherited *through* females (since Y does not pass through females).

The T-Chromosome in Drosophila

The functions of the Y- chromosome have been more thoroughly studied in Drosophila than in any other organism. Most of the facts as to this have already been brought out, in connection with our account of the functions of the X-chromosome.

We may summarize as follows the known facts as to the functions of the Y-chromosome in Drosophila:

- The Y-chromosome, although it normally occurs in males, is not required for the production of male individuals. Through non-disjunction, male individuals are produced that contain no Y.
- But the presence of Y is necessary for the fertility of males.
- Y may exist in females, as a result of accidental nondisjunction; it does not affect their sex.
- During many years of intensive study no characteristics were found that follow the Y-chromosome.
- But recently it has been shown that the presence of Y prevents the appearance of a certain recessive character ('bobbed') that is due to a modified X-chromosome. This modified X results in the production of shortened bristles on the body of the fly. But this peculiarity does not occur if a Y is present, so that Y has the dominant normal allele for 'bobbed'. Indications of certain other ill-defined functions of the Y-chromosome have of late come to view.
- No diverse modifications or defects such as are known in great number for the X-chromosome are known in the Y-chromosome.
- Thus in Drosophila the Y-chromosome appears to have little function; it seems nearly but not quite inert.

Active T-chromosomes in certain Organisms; Relation to Male Secondary Sex Characters

In certain organisms the Y-chromosome is more active and varied in its effects. In such cases it shows in different individuals diverse modifications, comparable to the diverse types of the X-chromosome in Drosophila, and these produce diverse inherited characteristics that appear in males only. This matter is closely bound up with the question of what produces the diversity of secondary sex characters. In many animals there are special characteristics which, like the Y- chromosome, are found only in males. These are the male secondary sex characters: horns, mane, beard, greater size, diversities in form, and the like. Since these appear only in the male, the question arises as to whether they may be due to the presence of the Y-chromosome.

With relation to this the following is the situation:

- In most cases, the presence of male secondary sex characters does not depend on the presence of the Y-chromosomes.

This is shown by the following:

— There are animals which have no Y-chromosome, yet have male secondary sex characters: for example, the dog, the cat, the horse; also various insects.

— In Drosophila, males that contain no Y (as a result of non-disjunction) have the typical male secondary sex characters.

— In some breeds of sheep the male has horns, while the female has none. It could be suggested that this is due to he fact that the male carries the Y-chromosome, while the female does not. But in other breeds, as Dorsets, both sexes have horns, although only the male has the T-chromosome. And in still other breeds, as Suffolks, neither sex has horns, although here as elsewhere the males have the Y-chromosome.

If a female of the horned breed (Dorset) is crossed with a male of the hornless breed (Suffolk), in the Fi descendants the males have horns, although they receive their Y-chromosome from the hornless breed. They have inherited the horns through the mother.

Thus it is clear that in these cases the presence of horns is not due to the presence of the Y-chromosome. Extensive work on the inheritance of horns and horn-lessness in sheep, goats and cattle consistently gives results which show that the presence of horns is not due to the presence of the Y-chromosome, or of any special type of Y-chromosome. In all these cases (a), (b) and (c) the presence of distinctive male secondary sex characters depends in some way on the fact that the male carries but one X-chromosome, and consequently produces the male hormone. Their absence from the female results from the fact that she carries two X-chromosomes, and consequently produces the female hormone.

- But in certain fish, some of the male secondary sex characters apparently do depend on the presence of a Y of a particular type, for they occur only where that type of Y is present, and they can therefore not be inherited through the mother. That is, the Y-chromosomes of certain individuals or breeds are so modified as to produce certain definite characteristics in the individuals that contain them, while individuals that contain another type of Y do not show these characteristics.

An example of this is found in the small fish Lebistes. In a certain race, the males have a black spot at the base of the dorsal fin, while the females are without the spot. In another race of the same species neither the males nor the females have the black spot. Thus the females of the two races are alike in lacking the black spot, while the males differ. Is this difference the result of differences in the Y-chromosomes of the two kinds of males? This may be tested by crossing members of the two races. In the sptted race, represent the X and Y-chromosome by capital letters; while in the non-spotted race, represent them by the small letters x and y.

A female of the spotted race, XX, is mated with a male of the non-spotted race, xy, thus:

- Females Males
- Parents XX xy
- Non-spotted
- *Fi* Xx Xy
- Non-spotted

The Y-Chromosomes

This produces in Fi females Xx, and males Xy. Neither contains the Y from the spotted race, and neither is spotted. The spot is not inherited through the females of the spotted race.

Next mate the female of the non-spotted race, xx, with a male from the spotted race, XY, as follows:

- Females Males
- Parents xx XY
- Spotted
- Fi xX xY
- Spotted

This yields in Fi females Xx, and males xY. These males carry the Y from the spotted race, and they are spotted. Thus far the spot goes with the Y from the spotted race. This Y may be traced further, for example, by breeding together the non-spotted females xx with the Fi males xY, from the preceding cross, thus:

- Females Males
- Parents x xY
- Spotted

- F2 xx xY
- Spotted

This gives in the Fs females xx, and males xY. These males are again spotted; they carry the Y from the spotted race. Wherever this Y from the spotted race is found, the black spot appears; wherever the Y is from the non-spotted race it does not appear. It appears clear therefore that whether the black spot appears in the male depends upon the type of Y- chromosome that it carries. There are two types of Y-chromo-some in the species, one producing the black spot, the other not producing it. It has been shown by Winge that there are many other programmes markings in this fish that depend for their occurrence on the presence of a particular type of Y-chromosome. They occur only in the males, and only in males that have a certain type of Y. In certain other fish also the Y-chromosomes have been shown to be diverse in different individuals, giving rise to distinctive characteristics in the male. In a certain plant, Melandrium, also, there are inherited characteristics that depend upon the presence of a particular type of Y-chromosome.

There are some indications also that in man the nature of the Y-chromosome affects certain characteristics. In a family described by Schofield, the male parent in the first observed generation had webbed toes. This was inherited in the male lines only by all the males for four successive generations. It thus followed exactly the distribution of the Y-chromosome from the original male parent. It seems probable therefore that it is due to a defect or modification in the Y-chromosome of that individual. Many other cases of webbed toes or fingers are known that occur in both sexes, so that they are clearly not due to the Y-chromosome. On the whole, the situation as to the Y-chromosome and its relation to inheritance may be summarized as follows:

- In most organisms Y is a degenerate chromosome, lacking most of the functions manifested by the other chromosomes.
- In some organisms degeneration has gone so far that the Y-chromosome has completely disappeared.
- In other cases it exists, but is very small.
- But in some species the Y-chromosome has effects on hereditary characters that are similar to those produced by other chromosomes. Different types of inherited characters result in such cases from diversely modified Y-chromosomes.
- Characteristics dependent on the presence of a particular type of Y-chromosome are inherited in the male line exclusively.

Three Main Methods of Distribution of Inherited Characteristics

There exist therefore three diverse systems of distributing hereditary characterstics, depending respectively on the X- chromosomes, the autosomes, and the Y-chromosomes. In each of these the rules of inheritance are given by the special method of distribution of the three types of chromosomes, taken in connection with the facts of dominance and recessiveness.

Inheritance which follows the distribution of the X-chromosomes is commonly called sex-linked inheritance. That following the distribution of autosomes is known as autosomal or typical Mendelian inheritance. Less important than those just named is Y-chromosome inheritance. All these methods are included under the term Mendelian inheritance, employed in a broad sense. In some organisms still other methods of inheritance exist. In higher animals and plants most inheritance occurs in one of the three methods named in the preceding paragraph.

Characteristics Affected both by Genes and by Environment

Many of the same features that can be altered by changing the genes can likewise be altered by appropriate changes in the environment. Characteristics do not fall into two classes, one exclusively hereditary (or dependent on genes), the other exclusively environmental; but any characteristic is affected both by the materials of which the organism is composed, and by the action of the conditions on these materials. Yet, as we shall see, in practice some characteristics are more readily altered by environmental conditions than are others. Certain characteristics owe most of their peculiarities to diversities among genes. Others are readily affected both by genes and by environment. Still others depend mainly on environmental conditions.

Particularly illuminating are the cases in which characteristics depend in marked degree both on genes and on environmental conditions, so that they are altered by changes in either. A number of well-known cases of this kind will be presented, selecting those that are from a technical point of view fully known. In that animal whose genetics is best known, the fruit-fly Drosophila, many of the characteristics that have been studied are defects or abnormalities which are sharply dependent on the presence of particular genes. Such a one is abnormal abdomen. This appears in the fact that the abdomen is ill formed, the segments not being regular nor sharply marked off. This abnormality is found to be due to a defect in a certain gene of the X-chromosome, so that it shows sex-linked inheritance, the abnormality being dominant. Unless that particular defective gene is present the abdomen is normal. But it is dependent too on the environment.

Individuals with the abnormal gene, if developed in a moist atmosphere, have the abnormality. But if those same individuals are developed in a dry atmosphere, they have normal abdomens. To produce the abnormal abdomen, we must have first the particular kind of defective gene that yields it. But second, even if that is present, we must have, too, the moist atmosphere for development; without it the abnormality is not produced, whatever the gene.

Another character in the fruit-fly that acts in a similar way is what is called reduplicated legs. If a certain defective gene is present in the X-chromosome, the animals show a tendency to produce double legs, or even triple or multiple legs, in place of single ones. But this does not happen unless the young develop at a low temperature. To produce reduplicated legs, then, there must be present a certain type of gene, and also a certain extenal condition. Unless both are present, the peculiarity is not produced. It is clear from these cases that even though an individual inherits the type of gene necessary to produce a certain defect, it may not be inevitable that he should have that defect. A special environment may correct what inheritance leaves imperfect These considerations apply to many thngs in man as well as in other organisms. Another example of the fact that what is produced in development depends both on th genes and on the environment is seen in the production of giants in the fruit-fly. A modified gene, located in the X-chromosome, near its left end, causes the animals (if they have no other type of X-chromo-somes than this) to become giants; they grow to nearly twice the size of the ordinary fly. But this increase in size takes place only if the animals having the modified gene are well fed during a certain period of their early lives. If not well fed at this particular time in life, they grow no larger than the usual flies. The giant size requires to produce it a particular type of environment acting on a particular type of gene. If either condition is not fulfilled, giants are not produced.

Again, in plants, the difference between green plants, containing chlorophyll, and white ones, is in some cases due to a difference in genes. Plants with a certain kind of genes will remain white even though grown in the light. In other cases, the difference between white and green plants is not the result of a difference in genes, but of a difference in environment. Plants that live in the dark remain white, even though they contain genes that can produce chlorophyll, while similar plants grown in the light are green. That is, to produce chlorophyll, a certain type of genes is necessary, and also a certain type of environment. Again, there are red and white varieties of primroses; the difference in programmes is inherited; it is due to gene diversities. But a certain variety produces red flowers when grown in a cool place, white flowers when grown in a warm, moist region, as in a greenhouse. The same diversity in programmes is now due to an environmental difference. Complex interrelations of genetic and environmental action have been described by Emerson with relation to diverse colours in maize plants. In different types, the leaves, the husks, the flowers, and other parts, differ in programmes. Emerson describes and Figs six main types, under each of which are subtypes.

These are the following:

- Purple, with its subtype, weak purple.
- Sun-red, with a subtype, weak sun-red.

- Dilute purple.
- Dilute sun-red.
- Brown.
- Green, with a number of subtypes.

The interrelations of genetic and environmental factors in these colours may be summarized as follows:

- Each programmes and each subtype is constant and hereditary, when the different types are bred under like conditions.
- Each of the colours depends on many different genes, like the eye programmes in Drosophila; they may be altered by changes in any of these genes. The different colours result from different combinations of the genes.
- When the different types are crossed they give typical Mendelian inheritance: in some cases with simple ratios, in others with complex ratios. Thus far the conditions are those for typical hereditary characters; the differences are the result of gene differences, and they follow the distribution of the different types of genes in inheritance. But, also, the colours depend on the conditions to which the plants are subjected while developing. Change of certain conditions changes the colours, as does alteration of the genes. Some of the relations are the following:
- The 'sun-red' type requires exposure to the sun for production of the red programmes. Parts protected from the sun are green. In the 'green types on the other hand, subjection to the sun does not cause the production of red programmes. Thus the difference between red and green plants is due in some cases to diversity of genes, as when 'sun-red and 'green types are both grown in the sun. In other cases the difference between red and green plants is due to environmental difference, as when 'sun-red plants grown in the sun are compared with those of the same type grown in the shade. The plants of the 'purple' type do not depend on exposure to the sun for their programmes; they are purplish in programmes whether grown in sun or shade. The degree of programmes produced by exposure to the sun depends on what genes are present. In the typical 'sun-red' variety, exposure to the sun causes a red programmes widely distributed on the plant. In the 'dilute sun-red', exposure to the sun under similar conditions causes only a little red, at the bottom and tips of the leaves, while in the 'green type no red appears. The nature of the soil affects the programmes. 'Dilute sun-red' grown in poor soil and subjected to the sun has almost all parts red; in good soil, with the same relation to the sun, it has very little red. The 'green' and 'brown' types are not influenced in their programmes by the nature of the soil.

- Certain particular chemicals in the soil affect the programmes. Presence of much nitrogenous matter tends to suppress the red; if little is present the red programmes is marked.
- Storage of carbohydrates in the leaves of the programmes-producing varieties causes increase of the red programmes. The amount of carbohydrate in the leaves may be increased by experimental procedures, as by removing the growing ears, or by bending a leaf so as to break the conducting vessels; the leaves thereupon become red in the parts with increased carbohydrate. But this effect is not produced in the 'green' types; they do not produce the red programmes even in presence of stored carbohydrates.

It appears clear, therefore, that in maize different genes cause the different varieties to have different metabolic processes, different chemical reactions, some of which cause the production of colours, while others do not. Also, diverse environmental conditions likewise cause changes in these metabolic processes, inducing programmes production in some cases.Whether programmes occurs, and its type and amount, depend on interaction of genetic and environmental conditions.

Another instructive case of the interaction of genes with environmental conditions is furnished by the conditions determining the number of facets in the compound eye of Drosophila. In the normal eye there are present about 800 facets. In certain of the flies a gene at locus 57 in the X-chromosome is so modified as to cause a great reduction in the number of fully developed facets. In this condition, known as Bar-eye, there are only about 80 fully developed facets: these form a 'bar' across the eye. This Bar-eye condition is dominant; the number of facets is reduced even though one of the two genes of the pair is normal. We may call the modified gene that causes the reduction in the number of facets the Bar gene.

The number of fully formed facets, it is found, depends on whether there is present but one or two of the Bar genes, and also on whether there is present in addition a normal gene. In the males (having of course but one X), the presence of the Bar gene reduces the number of normal facets to about 90. In the female, if both the X's contain a Bar gene, the number of facets is about 65. If the female is heterozygotic, so that but one of her X's contains the Bar gene, while the other is normal, the mean number of facets is about 358. Finally, in either male or female, if no Bar gene is present, the number of facets is about 800. Later it was discovered that this Bar gene is sometimes modified in a different way, so as to cause a less marked reduction in the number of facets. This modification is known as infra-Bar. A female containing but one of these infra-Bar genes has about 716 facets; if she contains two she has about 348.

Again, by an extraordinary process, almost or quite unique in known genetics, in some cases two of the Bar genes get united by crossing-over into one X-chromosome. This condition is known as double Bar. When both of the chromosomes carry this double Bar, the number of facets is but about 25. If one carries double Bar, the other single Bar, the number of facets is about 36. Thus the number of facets in the eye depends in many ways on the number and conditions of the genes present. This may be represented as: representing the conditions in the two X (or Y) chromosomes of any individual, with the corresponding numbers of eye facets present. In this diagram + = the normal, unmodified X-chromosome Y = the Y-chromosome of the male *B* = the Bar gene *iB* = the infra-Bar gene *BB* = the double Bar condition (two Bar genes in one chromosome) Ghr. *i*: + *iB B iB B iB B BB BB BB BB* Chr.s: + + + *iB* Y *BB* + *iB B BB* Facets 800 716 358 348 98 73 65 45 42 36 25. All these various conditions are inherited as unit-difference, sex-linked characters; they are typical hereditary conditions.

But it is found further that the number of facets in most of these conditions depends also on the temperature to which the developing animals are subjected. The numbers given above are those observed by investigators working with fruit-flies cultivated at ordinary, not precisely controlled, temperatures. If different sets of flies are kept at different temperatures, it is found that high temperatures reduce the number of facets in the individuals that carry the Bar genes, though not in the normal individuals. Krafka found the following mean numbers of facets in the various gene conditions, at different temperatures:

Males Degrees	*Bar Females B-B*	*Bar Males B-Y*	*Double Bar BB-BB*	*Double Bar BB-Y*	*Temperature,Females*
J5	214	270	5i	61	
20	122	161	33	37	
25	81	121	25′	28	
300131	40	74	‘5	‘4	

Thus the number of eye facets depends both on the genes and on the temperature. The number 61 may be produced either by a double Bar Male (*BB*–Y) at 15 degrees or by a Bar Female (*B*–*B*) at between 25 and 30 degrees. Changing either the genes or the temperature will change the number of facets.

Similar situations occur in vertebrates; one of these may be described briefly. The spotted salamander, Salamandra maculosa, is largely black in programmes, with spots or stripes of yellow. The amount of yellow differs in different cases, so that on the whole some specimens are lighter, some darker.

In the lighter individuals the yellow spots tend to run together into rows. The different grades are heritable; certain stocks are regularly darker, others lighter. If darker and lighter individuals are mated, the two conditions show Mendelian inheritance, the darker condition being dominant all the Fi generation are darker, while in Fa there are three dark to one light. The difference between the two thus depends on a difference in one gene. But it can be shown also that the extent of programmes depends on the environment. If the salamander is kept continuously for a long time (a year or two) on a light background, the yellow spots increase in size: they may run together into stripes or into large yellow areas *B*, so that the animals become much lighter. If, on the other hand, they are kept for long periods on a dark background the yellow areas decrease in size; the animals become on the whole darker. The extent of the yellow programmes thus depends both on the genes and on the environment; it can be altered by appropriately changing either.

In all these cases the following general principles apply:

- What are really inherited (passed on from parent to offspring) are the genes; that is, certain materials that in certain combinations and under certain conditions give rise to certain definite characteristics.
- With the same original genes, different environmental conditions may induce the production of diverse characteristics.
- Also, with the same environmental conditions, different genes may induce the production of different characteristics.
- The same difference in characteristics that is in one case produced by diversity of genes is in other cases produced by diversity of environment.

There can be little doubt that these general statements apply to many characteristics in man. Some combinations of human genes form a individual that is a much better culture medium for the bacteria of tuberculosis than are others. A person who gets such a combination of genes will develop tuberculosis, if he lives in a region in which the germs of that disease are abundant; while another person, with a different set of genes, will not be subject to tuberculosis, even though living under the same conditions. And the person with the genes that make him susceptible to tuberculosis will not have the disease if he prevents infection by the tubercle bacillus.

In relation to such susceptibility there are doubtless a great number of grades, dependent on just what combination of genes is present. Some are extremely susceptible, others less so but still prone to the disease, and so on, up to individuals whose gene combinations are such as to give them natural immunity to tuberculosis.

What then shall be answered when it is asked whether tuberculosis is hereditary or not? One can only say that some gene combinations predispose to it more than others, so that a hereditary factor is involved. But also there is a necessary environmental factor; it is the interrelation of the two sets of factors that determines whether the individual shall be afflicted with the disease. Similar considerations apply to any other disease, acute or chronic. Certain combinations of genes are doubtless more readily attacked by plague, by smallpox, by typhoid, by pneumonia, than others, just as certain genetic constitutions yield more readily to extremes of temperature, to exposure to the elements, to unfit food. There is no affair of life in which the gene combination borne by the individual does not play a part. But the environment too plays a part, often an overwhelmingly important part.

The relative role played by genes on the one hand, and by environment on the other, is not the same for all characteristics. Differences in certain features of the organism are mainly due to differences in the genes of which they are composed. In other characteristics, differences are due mainly to environmental diversities. There are all sorts of intermediate conditions between these extremes. We may pass in review a number of different types of characters from this point of view. Most of the diversities in obvious physical features in most animals are the result of differences in the genes, so that they are mainly matters of inheritance rather than of environment. In Drosophila the forms of the parts of the body, of the legs and wings, the distribution of bristles, the venation of the wings, the colours of body and eyes, are all changeable through the action of diverse genes, but diversities in these features due to environmental action are relatively rare. This is the case also for such characters in the vertebrates, including man. Most of the physical characters of human beings form of their features, complexion, programmes of hair, form of limbs, and the like, as well as sex are settled mainly while the child is carried in the mother's body, when the differences in environment are very slight.

The diversities between individuals in these respects are almost entirely though not quite entirely the result of differences in the genes; they are matters mainly of inheritance. This is demonstrated by what we find in twins. The one-egg twins, having the same genes, are usually closely alike in their physical peculiarities, while the twins derived from different eggs, and so having different genes, are much more diverse with respect to such physical characteristics. At times, of course, one-egg twins become diverse through accidents or diseases which affect one and not the other. But physical diversities in twins due to genetic differences are much more frequent than those due to environmental differences.

In some organisms the gross physical features are readily altered by environmental diversities, so that differences between individuals resulting from the different conditions under which they have lived are common. Such is the case with the habit of growth in higher plants; trees and other plants grown under different conditions are very different in form. In some of the lower animals, such as the hydroids, a similar situation is found.

It is obvious that the details of behaviour are in a high degree under the influence of environmental conditions. The particular acts performed at given moments in any given organism are determined almost entirely by the conditions under which it finds itself. Yet organisms of different genetic constitution behave very differently under effectively the same conditions, and the general pattern of the behaviour of organisms may be largely determined by the genetic constitution.

Genes and Environment in Relation to Mental Characteristics

The relative riddles of heredity and environment are of special interest in man, particularly in relation to the characteristics that influence behaviour. For the study of these matters man is the most favourable organism that exists, although for most other relations in genetics he is a singularly unfavourable object of study. In man it is possible to make detailed studies of temperament, mentality, character, a group of characteristics hardly open to examination in other organisms. There occur in man 'identical' or one-egg twins, individuals having the same set of genes, the same genetic constitution throughout. And for comparison with these there occur fraternal twins: two individuals of the same parentage, the same age and living under the same environment, but developed from two separately formed eggs and thus having to some extent different sets of genes. By comparing these two kinds of twins with one another and with unrelated individuals, opportunity is presented for comparisons of the results of similarities and diversities in genetic constitutions with those of similarities and diversities in environmental influences.

As seen earlier the twins derived from a single egg are as a rule very closely alike in most physical respects; it is this that has caused them to be called identical twins. In some cases, however, the two differ in certain physical characteristics. There appear to be two categories of causes for these differences. First, there are physical differences that result from the way in which the separation of the twins occurred. Separation of the single egg into two does not occur in the first stages of development, as by division of the one-cell stage into two. On the contrary, it occurs much later, when many cells are present and the differentiation of bodily parts has begun, and it occurs at different ages in different cases.

There is reason to believe that when an equal division occurs at a rather early stage, before the right and left halves of the body have begun to show their characteristic diversity, the two twins produced are closely alike in their physical features. At a later period the right and left halves of the embryo have begun to differ, in the way that results as a rule in the greater strength and development of the right hand and arm. At such a stage, the embryo may divide into right and left halves, with results as follows.

Of the two embryos so produced, one will have a right limb that is already somewhat advanced in development, while the left limb, formed at the plane of separation, must begin anew. The other embryo will be in the reverse situation; it will have a left limb already somewhat advanced in development, but the right limb, formed at the plane of separation, must begin anew.

In each case the more advanced limb retains its advantage and becomes the one that after birth is strongest and most used. The result necessarily is that one of the twins remains right-handed, while the other becomes left-handed. This situation is not uncommonly found in identical twins. Similar differences in other unsymmetrical parts may be produced in the same way: for example in the whorls of hair on the head. In such respects the two members of a pair often show 'mirror-imaging', presumably due in each case to the fact that the original embryo had already become a little unsymmetrical before division occurred.

From this same situation another condition often observed is believed to result. In the embryo before division, one side (usually the right) will be a little in advance of the other. After division, the twin derived from the right half continues to retain this advantage, so that it may be more vigourous than the one derived from the left half. Such differences in vigour are not rare in twins.

Apparently they may be accompanied by a psychological difference; the more vigourous twin assumes the leadership in their lives together. This reacts further on temperament and character. How far psychological and temperamental differences so produced may go is uncertain.

It appears that in some cases division of an embryo may be unequal, so that for this reason also one twin may be more vigourous than the other, with resulting psychological differences. Diversities produced in the ways above described are different in origin and type from those commonly called environmental. Under the latter term are commonly grouped diversities resulting from the different conditions under which individuals have lived. Since the extent to which differences due to the time and method of division may go, and the kind of characteristics that they may affect are uncertain, it is often not possible to distinguish these certainly from diversities due to differences of environment after division has occurred.

Belonging to one or the other class, there are sometimes marked differences in the phsyical characteristics of one-egg twins. Komai and Fukuoka have described two fifteen-year- old Japanese boys that are obviously one-egg twins; they show the usual close resemblance in form, features, colouring, and the like. But they now differ greatly in size.

One is taller by 14^{-8} centimetres, or about 6 inches, and his weight is greater by 10^{-4} kilograms, or about 23 pounds. The two were at first of the same size, the difference gradually arising during growth. The smaller of the twins is affected by diabetes insipidus; this is probably connected with his decreased growth. The larger twin is right-handed, the smaller left- handed. Whether thc difference in health and growth is in some way connected with what occurred at division of the egg is uncertain. A comparable case is described by Siemens. Two sisters, clearly one-egg twins, were as usual nearly identical in most of their physical characteristics till the age of 10. At this time one of them became affected wth a severe lateral curvature of the spine, and from this time their development was very different; at 16 the healthy twin was 4-8 inches taller than the other. The origin of the defect that produced curvature of the spine and changed development is unknown.

Identical twins that have lived together sometimes show considerable psychological differences. Newman has made extensive psychological tests on fifty pairs of identical or one-egg twins, and, for comparison with these, on fifty pairs of fraternal twins. The results of these examinations have not yet been published in full, but certain data from them have been published.

By the Stanford-Binet tests, the intelligence quotients (IQ) and the 'mental age' were determined for the two sets. The two classes of twins are best compared by the differences between the scores made by the two members of each pair. Thus the average difference between fraternal twins, having somewhat diverse genetic constitutions, is nearly twice as great as that between identical twins, in which the genetic constitutions are alike.

But it is important that the identical twins that have lived together also showed differences between members of the pairs. In different pairs there was much variation in the amount of diversity. Of the 50 pairs there were five in which the two members gave identical scores in the intelligence tests. At the other end of the scale were five pairs that showed respectively the following differences (points in the Stanford-Binet scale) between the two members: 12-6, 12-9, 13-0, 13-9, 16-0.

All these five differences are considerably greater than the average difference between the genetically unlike fraternal twins. It is clear that in some cases even identical twins living together differ much in mentality. Whether the differences are the result of original diversities consequent on the method of division of the egg, or have been produced in some way by different experiences of the two, is not known.

For comparisons as to the relative effects of genetic and environmental diversities on mental characteristics, valuable data have been obtained from the examination of identical twins that have lived apart under different environments. The first case of this kind was carefully studied by Muller, and nine pairs have since been fully examined by Newman. In each of these ten cases the twins were separated in infancy, being adopted into different families; they have then lived to adult life under diverse conditions and influences. In addition to careful observations as to character and temperament, they were subjected to psychological and temperamental tests by the best standardized methods available. The type of results reached by these studies will best be appreciated by first looking at the detailed resemblances and differences in some particular cases. In the first such study, by Muller, there were twin sisters that had been separated when two weeks old, and that did not see each other till they reached the age of 18; from that time until the age of 30 they lived apart more than nine-tenths of the time. Physically they showed the extreme similarity in characteristics that is usual in identical twins. Both 'have always been intellectually active', 'both have been extremely energetic, capable and popular, and they have been prominent in all sorts of club work in their respective communities' (Muller).

Both have had two or three attacks of tuberculosis, almost simultaneously'. The usual intelligence tests gave results very closely alike for the two twins. But 'the nonintellectual tests of motor reaction time, association time, "will temperament", emotions and social attitudes gave results in striking contrast with those of the intelligence tests, in that the twins gave markedly different scores in all these tests'. The differences were on the average greater than those between two individuals taken at random, and seemed 'to be correlated with salient differences in their past experiences and habits of life'. Thus this first study of such a case indicated that the different environments and experiences of the two individuals had produced a large effect on temperament, emotions, and social attitudes, but had had little effect on such matters as are tested by intelligence tests.

An illuminating contrast with these results is given by the first pair studied by Newman. The twin sisters ('O' and 'A') were born in London and were separated at the age of 18 months. One lived in Ontario, Canada, the other in London. Their environments were very different. When they were tested, this pair gave differences in those tests in which Muller's pair gave similarities, and gave similarities in those tests in which Muller's pair gave differences. Newman says: 'The twins dealt with in this paper are very different in mental capacity. But they showed great similarity in their manifestations of will and temperament, and in their emotional reactions. So this pair of twins shows that differences in the experiences undergone affect deeply the individual's mental traits: his performance in matters brought out by intelligence tests. Important points were brought out in the study of Newman's second pair, twin sisters. The two had received very different educations; one had attended school seven years longer than the other. Newman summarizes as follows the results of his study of these:

'These twins, remarkably similar after being separated at 18 months of age and unknown to each other for 19 years, have been profoundly modified by the very different educational careers.In every test of mental capacity, whether of so-called native ability or of achievement, 'G', the more highly educated twin, has distinctly the superior mind. Obviously mental training improves the ability of an individual to score well in any sort of test. But further, 'in contrast with the great difference in mental power stands the fact that in all the tests of emotional traits and of temperament the twins gave the impression of being remarkably and unusually similar'. Here came out the effect of their identity of genes.

Newman's third pair revealed certain other facts of importance. These were two young men ('G' and 'O'), separated at two months of age. One had lived mainly in the city while the other had lived in the country. They were examined after reaching the age of 23. In native ability they seem to be nearly identical. The one outstanding difference is in their general personalities. 'G' (who had lived in the city) impresses one as more dignified, more reserved, more self-contained, more unafraid, more experienced, and less friendly. He seldom smiles, has a more serious expression about the brows, eyes and mouth. He stands more erectly with chin held in and brows drawn down somewhat over his eyes. 'O' (the other twin) is the opposite in all these respects. He is the more typical country boy, laughs readily, and is not on his dignity at all. Newman asserts emphatically that the personalities of the boys were utterly different.

Thus through the study of these first four pairs examined, of identical twins that have lived apart, it became clear that such twins may differ in many ways. It appeared that the different environments and experiences of the individuals may have a large effect on mental and temperamental characteristics. The effects were different in ctifferent cases: in some cases the twins are alike in intelligence but differ temperamentally; in other cases they are alike in temperament and emotions, but differ in mentality. In other cases the marked difference is best expressed rather vaguely as a diversity in 'personality'.

The general impression given-by study of these first four cases has been confirmed by Newman's study of six other pairs of identical twins that have lived apart. The nine cases examined by Newman were treated by uniform methods, so that they may be compared in such a way as to bring out general relations. The mean difference in intelligence quotient for these nine pairs of identical twins that have lived apart is 8'6. This is considerably greater than the mean for identical twins that have lived together, which was 5-3. So far as it goes it indicates that difference of environment has a considerable effect in increasing the mental difference between twins.

In cases I, II, IV and VIII, the difference between the twins was very marked, but is similar to the differences between the five extreme cases of identical twins that had lived together. This emphasizes the fact that it is not possible to be certain what differences are due to something that happened at the division of the egg, and what are due to later environmental differences. The mean difference between identical twins that have lived apart is nearly the same as that for fraternal twins that have lived together.

In addition to the studies of intelligence, Muller and Newman made extensive examinations designed to test other features of personality. One set was designed to test will and temperament, to distinguish the mentally quick and slow, the deliberate and careful as compared with the hasty and careless, the aggressive and forceful or the opposite, and the like. Another type of examination is designed to test emotional peculiarities: likes and dislikes, feelings of right and wrong, and so on.

The results of such tests are not readily expressible by a numerical score, such as is employed for the intelligence tests. It will be worthwhile, however, to attempt a tabulation in general terms of the results of the tests in the three categories for the 10 cases of identical twins that have lived apart. Thus the general picture is that identical twins that have lived apart show many different combinations of likeness and unlikeness with respect to intelligence, temperament and emotional characteristics. There are cases in which the two are closely alike in all three categories cases in which the twins are distinctly diverse in all three categories and cases in which they are alike in two categories and diverse in the third, or alike in one category and diverse in the other two.

On the whole, it is clear that other things, beside genetic constitution, play important roles in determining mentality, temperament and emotional traits. The principal difficulty in interpreting the results lies in the uncertainty as to how much is the result of the manner in which division of the egg occurred, and how much is due to later environmental differences. Both certainly play a role. For these types of characteristics it appears probable that, of the two, the later environmental differences play the greater role. It is further clear that genetic resemblances and differences play a very large role in mentality, temperament and emotional traits.

It is to be noted that in all this study, the general environment of the twins was much alike in all cases. None of them lived in different civilizations, or at different epochs in cultural development. How great a diversity in the characteristics studied could be made by such differences remains uncertain. It is further to be noted that the tests to which the twins were subjected were designed to bring out rather permanent personal traits, as distinguished from mental content and habitudes. The behaviour of human beings depends very largely upon acquired mental content, upon knowledge, and upon habitudes and these things are in a high degree dependent on the conditions under which the individuals live and the experiences to which they are subjected.

In general, it is clear that in a human population there are great numbers of different types of individuals that are diverse because their genetic constitutions are diverse. In consequence of this they differ in their capabilities, in their tastes, in their tendencies towards any particular line of action. But all these classes, so far as they do not fall in the seriously defective groups, have marked powers of adjusting themselves to many different conditions and of following many different lines of action, depending on the conditions in which they live.

In no other organism than man is so full an analysis possible of the relative roles of genetic constitution and environment in determining the mental and temperamental characteristics, for in no other organism are these characteristics highly developed.

8

Genes and Adaptation

A species can adapt to a variable environment in one of three ways, two of which are genetic. It can retain a large reservoir of genetic variation, thus increasing the odds that under any stressful condition some members of the group are preadapted, and survive. In fact, most species in nature are genetically heterogeneous and the maintenance of the variability is assumed to be due to natural selection. Genetic uniformity is restricted to small, isolated, inbred populations. Another strategy is for a species to move in the direction of genetically programming a complex nervous system that modifies itself adaptively during development by responding to environmental stimuli.

The two modes of adaptation are not mutually exclusive and, in fact, often coexist in different proportions. We must be careful, however, not to infer genetic variability from behavioural variability, or vice versa. As an example, Ebert and Hyde selected for high and low aggression from a phenotypically intermediate wild population of mice. The range of phenotypes in the selected lines greatly exceeded that in the foundation population. Underneath the apparent phenotypic similarity of the wild mice was the genetic potential for extreme variation upward and downward. On the other hand, those of us who work with inbred strains of mice, each composed of genetically identical individuals, are often frustrated by the amount of behavioural variability within a single strain. Unfortunately for many experimenters, the behavioural phenotype is not rigidly predetermined by the genotype.

These two strategies of adaptation appear to be consequences of different situations. A species living in a stable, predictable environment will prosper best with a fixed behavioural repertoire that is precoded in its genotype as a plan for a specialized, unmodifiable nervous system. A species living in an unpredictable environment will do better to have its genotype programme the development of a plastic nervous system with the capacity for modification by appropriate stimuli. If part of the environment is predictable, a probably universal situation, selection will favour programming fixed responses for the predictable but not for the unpredictable components of its surroundings. A complete separation of these two modes of adaptation is impossible. Thus, modifiability of behaviour by experience along with genetic variation in its efficiency has been demonstrated in insects. And there appear to be biological constraints on learning even in such large brained creatures as humans.

Typical Behaviour

This discussion of the relations between behaviour genetics and ethology begins with two questions. How uniform is species typical behaviour? How much of its variation is attributable to genetic differences? In general, members of a species share a common ethogram as well as a common structural plan.

Even though taxonomists depend primarily on structure as a basis for classification, field observers know that behaviour is often a reliable means of distinguishing between physically similar species. The most striking examples are sibling species, particularly common in insects, fish, and amphibians although a few examples are known in birds and mammals. These species, however, are not our primary concern here.

There is ample evidence of behavioural polytypy within good species as defined by the free interbreeding of their members. Such polytypy is likely to be associated with social organization or food gathering, and explanations are frequently given in terms of ecological pressures and adaptive response.

Three major hypotheses could explain intraspecific behavioural heterogeneity:

- Variant behaviour is transmitted from adults to young by learning;
- Behavioural variation is an expression of genetic heterogeneity within the species; and
- Behavioural variation arises because individuals develop differently depending on the environment they encounter.

Hypothesis 1 is in a broad sense cultural, as it implies the handing down of traditional modes of behaving from generation to generation. Cultural transmission is nongenetic except to the extent that it requires a nervous system with a capacity for learning to model the behaviour of others. In humans, cultures are a major source of behaviour variation, although their prevalence within our species does not logically exclude a genic contribution to group differences.

A well known example of a group characteristic in animals is variation in food habits among troops of Japanese macaques. In one group the habit of washing food was apparently begun by one young female and was adopted quickly by her companions. It was not found in other troops living under similar circumstances. Cultural transmission of this type may be more important in nonhumans than is generally recognized, but because it has not been widely investigated from the genetic point of view we shall not consider it further.

Hypotheses 2 and 3 emphasize individualistic responses to environmental challenges rather than modelling behaviour on the actions of others. Culture is involved only as a part of the environment in which development takes place. The difference between the hypotheses lies in the role of genetic variation. Hypothesis 2 places a premium on genetic heterogeneity as a means of ensuring that some individuals will be able to survive and propagate in environments that deviate markedly from the species optimum. This is clearly the classical concept of natural selection, and it is supported in a general way by the universality of genetic variability in natural populations.

Hypothesis 3 does not deny the role of natural selection in the past, but for the present it emphasizes the lability of phenotypic development and adaptive responses to particular environmental conditions. For adherents of this view, behavioural variation it within a species is more a matter of differences in individual life histories than in individual genotypes. Actually, the three hypotheses are not mutually exclusive. All are supported by evidence, but their relative importance varies according to circumstances.

Behavioural Polytypy

An extensive literature deals with variation in the behaviour of domesticated and laboratory animals, house mice, rats, dogs, fowl, and fruit flies. In these species it is often possible to separate experimentally the effects of genetic and environmental factors. Genetic variation has been demonstrated for activity level, emotional reactivity, eating and drinking, learning ability, and social behaviour. In some domestic species, selection for behaviour has produced breeds with special aptitude for certain tasks. These breed differences persist even when animals are reared as nearly as possible in the same manner.

As an example, Fuller trained five breeds of dogs to walk on a lead from their outdoor living quarters to a laboratory and found striking differences in their reactions. Shetland sheep dogs crowded against their guides, a pattern that strongly resembled the herding of sheep by their working relatives. Basenjis, with a short history of domestication, struggled more violently and bit their leash more frequently than any other breed.

Beagles, who have been selected to track small game and to give voice as they follow a trail, were noteworthy for their mournful vocalization. None of these behaviours was restricted to a single breed, and none was completely absent from any breed. Their frequencies and intensities differed enough, however, to define significantly different breed profiles. Crosses between Basenjis and cocker spaniels, breeds with widely differing profiles, yielded hybrids that were intermediate to their parents on all characteristics except vocalization, which was higher than that of either parent. The occurence of stable, intraspecific behavioural variation among populations is ethological polytypy. It is not restricted to domesticated animals. Polytypy has been reported within single taxa whose range extends over a wide territory that covers different environmental conditions. Examples include baboons (Papio anubis), vermets (Cercopithecus aethiops), and langurs (Presbytis entellus), whose social organization varies greatly over their ranges and appears to be correlated with the nature of the habitat.

Hendrichs has recorded variations in the social organization of three species of antelope, diddik (Madoga kirki), reedbuck (Redunca redunca), and waterbuck (Kobus ellipsiprymnus), that are associated with population density, that is, in turn, presumably regulated by the supply of food and other resources. Stacey and Bock report that acorn woodpeckers (Melanerpes formicovorus) that ordinarily live in permanent groups and defend year round territories deviated from this pattern in one location. Here, they mated in pairs and migrated in early winter. Old field mice (Peromyscus polionotus) from different geographical areas were found to have identical patterns of sexual behaviour, but these varied in quantitative aspects such as latency, frequency, and duration of the components that make up the total mating act.

There is no evidence that these polytypies are attributable to genetic heterogeneity. In fact, many of the cited authors generally ascribe these variations in social organization and behaviour to environmental pressures. The assumption, not always explicitly stated, is that given the same circumstances any member of the species could adopt either of the alternate patterns. However, there have been few tests to disprove a genetic basis, and it would be difficult to make one with many of these species. It is possible in species that can be maintained in large numbers in captivity and bred systematically.

Important contributions to the genetic analysis of ethological polytypy have been made by King and his students. The genus *Peromyscus* includes many taxa that have been classified at the subgeneric, species group, species, and subspecies level in ascending order of closeness of relationship. King, Price, and Webber compared animals from nine taxa on five motor behaviours: swimming, running, climbing, digging, and gnawing. They were searching for correlations between behavioural similarities and closeness of relationship as judged by taxonomists.

Secondarily, they also looked for correlations between behaviour and habitat. For example, burrowing species might be expected to excel in digging, woodland species in climbing. In all tests the only motivation to perform any of the actions was to escape from an unfamiliar place. Differences among taxa were found on every task, but the expectation that the degree of similarity would correlate with the degree of relationship was not fulfilled. Surprisingly, the behaviour—habitat correlations were no better.

Although the results of this study were disappointing, King rethought the problem and carried out other experiments. Instead of comparing taxa on specific motor acts in an artificial setting he designed a set of tests to measure more general behavioural characteristics: speed of adaptation, reflex responses, abstractive and strategic behaviour. Again the subjects were *Peromyscus* taxa of varied degrees of relationship and similarity of habitat. The results of this study lead to a revised view of behavioural polytypy as related to genetics. All taxa of *Peromyscus,* and many other "mice," are roughly similar in form and are capable of "rodentlike behaviour" such as gnawing, grooming, and nest building. The major differences among them lie in their modes of organization of activities in new situations as they search for food, mates, or shelter. Some taxa become specialized through fixation of genes favouring survival under typical natural conditions; others retain their juvenile plasticity into adulthood and adapt as best they can to situations as they arise. King finds evidence that both Hypothesis 2 (fixed genetic programming) and Hypothesis 3 (flexible neural programming) apply to *Peromyscus.*

In a somewhat similar experiment with wild *Mus musculus,* captured in three ecologically distinct areas and reared similarly in a laboratory, Plomin and Man osevitz found differences in open field activity, defecation, climbing, running wheel activity, and nest building. They attributed the differences to genetic factors, although controls for maternal effects and cross breeding experiments were not included in their study. A repetition with more subjects and additional controls would be useful.

In muroid rodents, male copulatory behaviour has been studied from a genetic point of view. Dewsbury (1979b) reported basically similar factor patterns in laboratory rats, the descendants of a wild population of house mice, and deer mice. In spite of the factorial similarity each of these species has a distinctive style of courtship. In more closely related taxa, such as subspecies, male copulatory behaviour is much more uniform and for practical purposes qualitatively identical. However, significant quantitative differences in the latencies and durations of phases of mating were found in 11 of 14 measures. Subjects were three subspecies of *Peromyscus maniculatus: P.m. bairdii, P.m. blandus,* and *P.m. gambeli.* Cross fostering of deer mice on house mice produced minor quantitative effects on latencies, but, according to Dewsbury: "basic motor patterns in cross fostered mice were identical to those of normally reared animals." A particularly interesting instance of behavioural polytypy occurs in the squirrel monkey (*Saimiri sciurus*). Two populations of this species differ markedly in the phonic characteristics of their extensive vocabulary of calls. Externally the populations are distinguishable by facial markings. The Roman arch type found in Peru, Costa Rica, and Panama, and the Gothic arch type resident in Columbia and Guyana also differ in their karyotypes.

That the dialects differ for genetic rather than cultural reasons was demonstrated by rearing infants with devocalized mothers and finding that they conversed in the dialect appropriate to their chromosomes and emitted these calls in their appropriate context. A personal communication from Newman adds that new research has shown that isolation peeps of these two forms differ so much that analysis of sonograms permits classification of individuals with 99 per cent accuracy. The recorded IPs of eight F_1 hybrids were matched with their phenotypic appearance. Five hybrids with Roman arches emitted typical Roman IPs. The IPs of the three Gothic type hybrids were variable, one strongly Gothic, one strongly Roman, and one mixed. These observations are insufficient to hypothesize a mode of inheritance, but apparently vocalization patterns are associated to a degree with physical characteristics.

The Traditional Genetic Procedures of Crossbreeding and Selection

Most knowledge of the formal genetics of social behaviour comes from experiments with selected breeds or inbred lines of house mice, Norway rats, dogs, *Drosophila*. Can results obtained with these long domesticated animals lead to principles that can be extended to their wild relatives?

Lorenz asserted that results from animal strains that have been bred for a long time in captivity are not applicable to free living animals: "because captivity changes all hitherto effective selective factors in so profound a way that serious changes must be expected in the genome of the stock after only a few generations". Recent research indicates that Lorenz was too sweeping in this statement. Domestic albino rats transferred from their protective laboratory environment to outdoor pens survived severe northern winters, maintained a stable population, constructed and lived in burrows indistinguishable from those of wild rats, and in general behaved like their wild cousins.

Given their freedom, the albinos were able to cope with stresses that their recent ancestors had never encountered. Of course, in a completely feral state their impaired vision and conspicuous colouration would make them more vulnerable to predators. The important point of Boice's study is that there was no evidence of a loss of species typical adaptive behaviour. Disuse of skills does not by itself lead to their deletion from the behavioural repertoire. In a similar vein, Miller made extensive observations on Peking ducks that have been separated from their wild mallard ancestors for centuries. Pekings differ from mallards in important characteristics: They are too heavy to fly, and they lack the mallard's striking sexual dimorphism in plumage. Yet, when observed in a natural setting during three breeding seasons, they performed all the species typical displays that have been reported for mallards.

Miller concluded that his data: "offer no justification for the view that domestication has a 'degenerative' effect on social courtship behaviour patterns, but are rather supportive of the view that the customary nonoccurence of these displays is a function of the inhibitory or unfavourable environmental context in which the domestic birds usually find themselves." The observations of Boice and Miller do not negate the existence of behavioural differences between domesticated animals and their wild progenitors but they support the idea that many such differences are the results of selection for a characteristic desired by their breeders: docility in laboratory rats, and flightlessness in birds that are intended for the dinner of their custodians.

Some changes accompanying domestication probably are the effects of the loss of dominant alleles through inbreeding. Lynch found that inbred wild house mice, compared with random bred animals from the same foundation stock, had smaller litters, bore lighter offspring, and constructed poorer nests. Random bred wild mice reared in a laboratory were not larger, more fertile, or better nest builders than standard inbred strains. In fact, their nests were smaller than those of the heterogeneous HS stock produced by first intercrossing eight inbred strains and then continuing with a random breeding system. Lynch explains the hypertrophy of a behavioural trait related to fitness in a stock synthesized from inbred strains that built only normal sized nests by postulating that some deleterious recessive alleles present in wild mice were eliminated during inbreeding. She believes that the inbred lines that survive the inbreeding process are good material for analysing the genetic architecture of a species.

Her view is contrary to that of Smith who found that the heritability of shuttle avoidance learning was much lower in lines of mice selected from a wild stock than in inbred strains and heterogeneous stocks. The relevance of shuttle box avoidance to biological fitness. There is considerable evidence that good performance in the apparatus depends less on learning ability than on the prepotent response to alarm signal, freezing or running. Domestic rats have been selected for docility, and this may indeed disqualify them for inferring selection pressures in nature that are related to escape from shock. Some of the changes in behaviour that accompany domestication seem to be the effects of inbreeding rather than of conscious or unconscious selection. Connor compared wild house mice reared in a simulated natural environment with laboratory reared wild mice and with three standard inbred strains. Wild mice, irrespective of their habitats, were more aggressive towards intruders, vocalized more when handled, and were more difficult to recapture when they were freed in a room.

In a second experiment, Connor followed behavioural changes over 10 generations in random bred wild mice maintained in a natural or a laboratory habitat. A third group from the same original population were inbred by brother sister matings. No behavioural differences were found between the laboratory and the seminaturally reared random bred stocks. Maternal effects were looked for in a cross fostering study, but none were found. In contrast, the inbred wild lines were less aggressive and easier to recapture. The evidence is convincing that we should not ascribe intraspecific variation in behaviour exclusively to genetic or to environmental factors without evidence from experiments designed to separate their effects. Price, on the basis of his studies of the modification of activity by mode of rearing in wild and semi-domesticated *Peromyscus maniculatus bairdii* came to the following conclusions.

The genetic requirements for a behavioural character should not change during breeding in captivity unless:

- It is selected for deliberately; or
- It is selected for unwittingly because it improves fitness under the conditions of captivity. These conclusions are encouraging. By a judicious mixture of field observations with controlled breeding and rearing in captivity, we someday may be able to determine the extent to which natural variability in species typical behaviour is a function of genetic differences. An example is the work of Moss and Watson, *Sociobiology*. Genetic heterogeneity is an absolute necessity for evolution; we know that it exists, but we do not know how important it is with respect to differences in behavioural fitness.

Genotype Environment Interaction

Neither genotype nor environment is sufficient to determine a behavioural phenotype. Without a genotype there is no organism. For a zygote or parthenogenetic ovum to develop into an adult there must be an appropriate physical and biological environment. Ethologists are especially interested in the biosocial factors that shape the development of behaviour. Indeed, the developmental process is often described as a genotype—environment (GE) interaction. The general phenomenon of joint involvement as GE coaction, and to confine GE interaction to situations in which a specific environmental change affects organisms differently depending on their genotypes. GE interactions are detected by exposing two or more genotypes to two or more conditions (*e.g.*, stimulated versus nonstimulated) and comparing the behavioural effects.

Broadhurst and Jinks distinguished two ways in which genes might affect development:

- They could influence its stability, particularly during early stages, and effect processes that could potentially influence an individual's behaviour for life. A tendency for phenotypic development to proceed on a stable, species-specific course regardless of ambient conditions is called developmental homeostasis.
- At any stage of life, genes and their productions may influence the capacity of an organism to adjust to changes in its environment such as removal from a familiar habitat. Broadhurst and Jinks' distinction between stability and change is somewhat arbitrary, but it is still useful. In general, stability refers to processes during early development that have generalized, permanent effects; change involves behavioural adaptations to a specific situation, with less carry over to other conditions.

GE interaction takes several forms. The most extreme is disordinal interaction where the effects of an environmental change are opposite in direction depending on the genotype of the subjects. Demonstrates that in a 2-genotype-2-treatment experiment, one could come to completely different conclusions regarding interaction depending on one's choice of treatments. An organism must receive an appropriate exposure to some stimuli in order to develop optimally. The low end of the environmental scale corresponds to extreme stimulus deprivation; the upper end to severe stress. Strains A and B differ in their optimal stimulus requirements.

A requires less than B, but it is also more easily damaged by high levels. Suppose now that we conduct four experiments with these strains choosing only two values of stimulus intensity for each. The chosen values are indicated by x's on the abscissa. Experiment 1 yields a clear interaction; strain A is helped by stimulation, strain B scarcely at all. Experiment 2 produces a marked disordinal interaction. Strain A is impaired as stimulation increases; strain B is improved. In Experiment 3, both strains profit from additional stimulation, and there is no interaction. Finally, in Experiment 4 both strains are impaired as stimulation increases.

Thus the nature of any interaction found may be a matter of the choice of experimental procedures rather than evidence for fundamentally different ways of responding to a stimulus change. The remedy, obviously, is to employ a wide range of treatments in any experiment where GE interactions are likely to occur. Unfortunately, there are few experiments extensive enough to provide a test of the generality of the threshold hypothesis. They were placed separately in cages where they had no physical or visual contact with other dogs or humans. Generally the period of isolation was 12 weeks; in some experiments isolation was interrupted periodically by brief placement in an arena where subjects could interact successively with a human handler, with another puppy of the same age, and with a rubber ball or swinging toy.

From the results of a series of experiments it was concluded that the deleterious effects of early isolation are primarily attributable to an overwhelming fear reaction when an adolescent puppy suddenly faces a complex environment that is totally unfamiliar. Isolated puppies that were sedated with drugs at the time of first emergence from isolation performed almost like normally reared individuals, demonstrating that their motor and perceptual development had been little impaired by living in the small chambers.

In most mammals reactive capacities develop gradually from infancy to maturity and the environment that must be dealt with at different ages is matched with the infant's or adolescent's perceptual and motor capacities. Rather small interruptions in isolation were sufficient to counteract the effects of experiential deprivation. Puppies who had 15 minutes per week in our test arena developed social and manipulative skills almost as well as pet reared animals with several hundredfold more experience.

A number of our experiments were directed towards defining breed differences in the intensity and duration of the effects of experiential deprivation with beagles and wirehaired fox terriers as the subjects. Beagles have a history of selection for olfactory tracking of small game animals. Litters housed together in group pens seldom fought and dominance hierarchies within litters were weak. In a familiar laboratory setting they gave the impression of relaxation, but they were somewhat shy in unfamiliar surroundings. Housed in the same way, the terriers engaged in fierce battles and formed strong dominance hierarchies within litters. When handled in tests their muscle tension was high, and they were more likely to attack than to retreat from a threat. Thus, it was not surprising that the effects of isolation were more striking in beagles than in terriers. In one experiment, beagles, transferred to the test arena in a small cage, often failed to explore the new environment; terriers rapidly overcame an initial hesitation and reacted appropriately to objects, people, and other puppies. In another set of experiments, beagles and terriers who had been isolated from weaning to 15 weeks of age were compared with pet reared siblings. Over a period of 5 weeks each subject was observed 25 times in the arena and records made of behaviours such as responding to a handler's call, following a moving person, playing with toys, and interacting with another puppy of the same age.

A quantitative scoring system was devised for two aspects of behaviour, response intensity to key stimuli and activity index. The response indices of pet reared (control) beagles and terriers were practically identical. In both breeds, isolates were inhibited in responding during early tests. Intensity scores rose with repeated exposure to the arena; the rate of increase was much higher in terriers. At the end of the study isolated beagles were still inhibited compared with controls.

On the second measure pet reared terriers were more active than similarly treated beagles throughout the observations. During early tests, isolates of both breeds were inactive, as had been anticipated from previous experiments and, as expected, activity increased in both breeds with repeated trials. Beagles were approximately equal with their controls by the fifth week. Terriers, however, were significantly more active than their controls by the second week.

At first glance this might appear to be a disordinal interaction. Experiential deprivation makes terriers more active; it makes beagles less so. In terms of overt behaviour this is certainly true. But there are other ways of looking at the data. In both breeds isolation produced an initial aversion to the arena, reduced activity, and decreased the intensity of response to the test stimuli. Both breeds recovered with experience, but the terriers improved more rapidly and more completely. To be sure, the isolated terriers overshot and became more active than their controls; isolated beagles were less active than their controls. In a sense, however, isolation acted similarly on both breeds in exaggerating the characteristics that differentiated them when they were raised conventionally. Normally active and investigative terriers became more so. Deliberate and cautious beagles became hesitant and timid. It appeared that isolation had made both breeds more emotionally reactive, but they expressed the change in divergent genotype typical patterns.

Genetic Architecture and Behavioural Phylogeny

Ethology is grounded in evolutionary concepts. Roe and Simpson's *Behaviour and Evolution* brought together the views of many distinguished scientists and is still an excellent introduction to behavioural phylogeny. The value of the comparative method in deducing the phylogeny of innate behaviour patterns, particularly communication by sound, posture, and gesture, was emphasized earlier by Lorenz.

However, there are problems with the application of evolutionary theory to behaviour. The validity of the very idea of behavioural phylogeny has been criticized on the ground that it is impossible to distinguish between similarity due to common ancestry (homology) and that attributable to common selective pressures (analogy). His warning is appropriate, but it need not lead to the abandonment of the consideration of the role of behaviour in evolution. In this presentation we are not concerned as much with homology and phylogeny as with the role of behaviour as a component of fitness and, therefore, as a directive force in evolution. Consideration of this topic requires a brief outline of theories relating the genetic architecture underlying variation in a behavioural character to the character's role in natural selection.

Mendelian Approaches to Behaviour Adaptation

We can distinguish between approaches based on one or two loci (Mendelian analysis) and those based on biometrical analysis of quantitative variability. The latter type of variation is considered to depend basically on Mendelian principles, but the numerous loci that are involved are not individually identifiable. Mather called these polygenes and we now apply the term polygenic inheritance to many behavioural traits. Both the Mendelian and the biometric approaches are important to behaviour genetics. Benzer's studies of mutants with well defined behavioural peculiarities have advanced our knowledge of neural development and function in *Drosophila*. The neurological mutants of mice have contributed to our understanding of brain development and function in mammals. Coat colour mutants, such as albinos, have been studied frequently by behaviour geneticists.

Albinos do differ from pigmented animals in many aspects of behaviour, and even normally pigmented carriers of the albino allele are impaired in some phases of development. However, neurological mutants and albinos are so rare in natural populations that they cannot be classified as true polymorphisms. We know that genetic polymorphisms are extremely common and the possibility that some of them have behavioural significance is well worth investigating.

An individual, the target of selection, is not a mosaic of characters each of which is the product of a given gene. Rather, genes are merely the units of the genetic programme that govern the complicated process of development, ultimately resulting in the phenotypic character. To consider genes as independent units is meaningless from the evolutionary point of view because the individual as a whole is the unit of selection. To regard genes as independent units is meaningless from the physiological viewpoint because genes interact with each other in producing the various components of the genotype.

For Mayr, the genotype has properties not deducible from the sum of its components. But he also acknowledges the possibility of switch genes where an allelic substitution at a single locus may shift the phenotype so that the direction and intensity of selection pressure is modified. A familiar example is industrial melanism, the replacement of light coloured variants of several species of moths by darker variants in areas where pollution has altered the colour of the bark of the trees on which the moths rest.

How much evidence is there for switch genes that change behaviour drastically enough to influence evolutionary processes? Sociobiologists seem to assume that such genes are abundant enough to be important. Trivers writes: "Assume that the altruistic behaviour of an altruist is controlled by an allele (dominant or recessive) a_2, at a given locus, and that there is only one alternative allele, a_1, at that locus and that it does not lead to altruistic behaviour." Simple allelic substitution could switch such complex behaviours as altruism and selfishness or change the mating system of a species from polygamy to monogamy. It is hard to imagine how such a switch gene would operate physiologically.

Of course, any allelic substitution that affects sensory, neural, or motor functions will have some effect on behaviour, but a single mutation seems unlikely, by itself, to shift behaviour into promising new channels. However, there are reports suggesting that some genetic polymorphisms have behavioural correlates that affect selective processes indirectly.

Krebs, Gaines, Keller, Myers, and Tamarin found large fluctuations in the frequency of alleles at the transferrin (*Tf*) and leucine aminopeptidase (*LAP*) loci in the vole, *Microtus pennsylvanicus*, that were associated with changes in population density. They believe that the gene frequency changes over time were indicative of an association between the alleles at these loci and the ability to survive and propagate in stressful situations. Specifically, they hypothesized that females heterozygous at the *Tf* locus were apt to migrate during periods of increasing population density. Homozygotes, particularly for the Tf^{c} allele tended to remain in their original territory. The implication is that the *Tf* locus has somehing to do with social interactions, but, to my knowledge, this relationship has not been verified by direct observation of individual encounters. Krebs et al.'s data were obtained from field experiments supplemented by genotype determination of captured animals. A group of immuno-geneticists have applied a very different approach to the genetics of social behaviour. Andrews and Boyse claim to have mapped two loci related to mating preference in house mice. They hypothesize that one of these, *Ris* (recognition of identity signal) provides a cue, probably olfactory, by which males can distinguish preferred females. *Ris* is assigned a location in the *Qa-Tla* region of the major histocompatibility region of the mouse. Another locus, Rir (recognition of identity response) is assumed to determine whether male response is positive or negative.

The genetic procedures are complex and ingenious, Males were given five to six repeated tests to determine their sexual preference. The significance of deviations from the null hypothesis (random choice) was apparently computed on the basis of the total number of tests rather than on the number of individuals.

The receptivity of the rejected females was confirmed by pairing them with other males, but one cannot exclude thepossibility that factors other than identity recognition by males bias the possibility of a female being bred. In spite of these experimental problems, the hypothesis that identity recognition based on cellular characteristics is important in mating choice has evolutionary implications. More strinent statistical analysis and more adequate behavioural controls are needed.

Isolation between closely related taxa need not imply a behavioural barrier. An example that involves a physiological difference with a relatively simple genetic basis has been reported in the lacewing flies, Chrysopa cornea and C. downsei. In the laboratory these species interbreed readily. In nature, crossing is not observed. C. carnea produces three generations per summer; C. downsei produces one, then enters diapause. Standard Mendelian crosses were made between the two species and the occurence of diapause in the hydrids was observed. The data are consistent with control of diapause by two loci. The C. downsei phenotype (diapause) requires recessive allele homozygosity at both loci. This case should probably not be called behavioural isolation in the strict sense, as the two species will mate when they are artificially manipulated. It does demonstrate that relatively simple genetic systems may have substantial evolutionary consequences.

Quantitative Inheritance

Let us now consider deductions that can be made regarding behavioural evolution from knowledge of the genetic architecture that regulated quantitative characteristics. It is well known that variation in such traits is largely polygenic. The theory relating genetic architecture to the evolution of behaviour has been explicated by several investigators. Essentially, stabilizing selection for an intermediate expression of a trait, and directional selection for high or low values involve different genetic strategies. Stabilizing selection requires either additivity of gene effects, so that a heterozygote is midway between the two homozygotes; or a balance between dominant genes at different loci, some with increasing, some with decreasing action on the phenotype. Traits with such a history respond well to selection in the laboratory and are said to be highly heritable.In contrast, selection for the extreme manifestations of a trait favours alleles with dominant effects in one direction only. Such a system ensures that the favoured alleles will be effective in a single as well as in a double dose. Over time, recessive alleles will tend to disappear from the selected population, and the heritability of the trait will become low. Clearly, it is most efficient for a species to have those traits most directly related to fitness regulated by dominant genes. Falconer noted that in domestic animals heritability is lower for traits such as fertility and viability, than for traits desired by breeders for purely economic considerations.

Accepting these principles, one can work backward from determination of the genetic architecture of a trait to a reconstruction of the role of that trait in natural selection. High unidirectional dominance is evidence that extreme values of that trait increase fitness significantly. Balanced bidirectional dominance implies a history of selection for intermediate values of the trait. High additive variance indicates that a trait has not been important for fitness, or that selective pressures are inconsistent and that the ability to vary phenotypes rapidly in response to a change in environment is valuable. There is an increasing number of studies in which the genetic architecture of a behavioural characteristic in domesticated strains has been used as the basis for inferences regarding the relation of that trait to fitness. Bruell demonstrated that F_1 hybrids between inbred strains of mice almost always surpassed both parental strains on activity tests.

The poorer performance of the parents was attributed to the inevitable, random loss of some dominant alleles during inbreeding. These losses generally involve different loci in any two strains; thus, in their hybrid offspring the deficiency of one is compensated by the contribution of the other. As might be expected, Bruell found that heterosis was more pronounced in crosses whose parents were distantly related than in crosses between lines with similar origins. More sophisticated modes of genetic analysis were used by Broadhurst and Jinks, who measured open field activity and defecation in six lines of rats and all possible F_1 hybrids. This diallel cross design provides estimates of many genetic parameters.

Their genetic findings were:

- Genes enhancing developmental phenostability tend to be dominant;
- Dominance also occurs for rapid decrease of emotional defecation on successive tests in the open field;
- Exploratory activity is also regulated by dominant genes, but these may have either a negative or a positive sign. From thse observations they concluded that selection has operated to:
- Buffer the fetal and infant rat from external stress;
- Promote a rapid return to normal emotional status after arousal;
- Strike a balance between bold curiosity that could lead to disaster, and timidity that could cause exclusion from essential resources. None of this is surprising, but it is satisfying to find that the genetic findings are consonant with our expectations of what organisms should do in order to prosper.

The genetic approach to reconstruction of evolutionary history is gaining ground, but there is difference of opinion regarding the best way to proceed. For example, Lynch concluded that: "because inbred mice are not weakling, analysis of crosses between them may represent a more efficient method for detecting genetic architecture than inbreeding of wild populations." Smith disagrees on the basis of his comparison of the heritability of shuttle avoidance in selected wild mice and in inbred strains.

He doubts that it is safe to deduce the selection history of a hypothetical parent popuation from the genetic architecture of contemporary inbred strains of largely unknown origin. Obviously, more research is needed to settle this matter. Henderson made an effort to escape the bind of first accepting the basic hypothesis of th relation between genetic dominance and fitness of a trait, then conducting a genetic experiment and, from its results deducing the fitness value of that trait. He deduced from observation that a particular trait was related to fitness and then conducted a breeding experiment to determine its genetic architecture. Specifically, he reasoned that mice should be selected for low levels of motor activity during infancy because they are safest when they remain in their nest. If they are accidentally dislodged their chances of being retrieved are greater if they remain quiet and await their mother. He predicted that, in direct contradiction to the situation in adults, low activity would be dominant in infant mice.

A triple test cross confirmed his prediction; the heritability of activity was low and inactivity was dominant. This was attributed by Henderson to a relaxation of selection for infant immobility over many generations of laboratory rearing. In a very different species, Cheng, Shoffner, Phillips, and Shapiro studied imprinting to visual and auditory stimuli in five groups of mallards (Anas platyrhyncs) ranging from a long established game farm line to 25%, 50%, 75%, and 100% wild ancestry. Wild ducklings imprinted better than domesticated ducklings, and heterosis was found in the hybrids, as would be predicted from the hypothesis that traits with adaptive value show little additive variation. Efficient imprinting seemed to be correlated with differences in arousal during early exposure to stimulation.

Confirmations of a prediction do not prove the universal value of a method. However the approach of Henderson and Chang et al. is promising and has considerable potential. As Henderon states, the method is most powerful when predictions can be made that a particular kind of behaviour will differ in its contribution to fitness at different parts of its life-span, or in dissimilar environments. The method is obviously limited to species in which multiple distinct lines or strains are available. Principles derived from such studies are, however, likely to have general application. It is only fair to state that the usefulness of this approach to defining adaptive behaviour has been challenged on the basis that the experimental subjects available for analysis of genetic architecture are not representative of natural populations, and that a history of inbreeding distorts the genetic structure typical of a species under natural selection. These warnings must be kept in mind but much of the data obtained in this way seems to substantiate accepted ideas of the characteristics of fitness enhancing behaviour. There is, of course, a possible circularity in such experimens.

One can hypothesize thats high levels of behaviour B_1 are adaptive and thus should have increased fitness during recent evolutionary history. Suppose now that an experiment is performed and low levels of B_1 are found to be inherited in dominant fashion. One could conclude that the original hypothesis was wrong, and that actually low levels of B_1 enhance fitness.

Alternatively one might consider the possibility that B_1 has little to do with fitness but, in the population studied, it is correlated genetically with another behaviour, B_2, that has been selected on the basis of enhancing fitness. The best way out of such circularity is good ethological studies relating B_1 and B_2 independently to survival and propagation. Another problem with the attempt to deduce the selective importance of a particular form of behaviour is the choice of the method of measuring it. Behaviour geneticists have been sure for years that in an open field C_{57}BL mice were more active than BALB/c mice. Whitford and Sipf found no important differences between these strains when the height of their open field was reduced from 36 inches to 1 inch by insertion of a plxiglass sheet. It is clear that open top, open field activity is very different from contact top, open field activity.

Plant Genetic Diversity

The conservation of plant genetic resources has, in recent years, attracted growing public and scientific interest and political support. There is an increasing awareness of the relevance of biological diversity and its conservation to the health of the biosphere. Particularly urgent is the need to raise agricultural output to meet the basic nutritional needs of increasing populations. The dilemma is that the required increase in food production must be obtained through 'sustainable' forms of agriculture which are less dependent on the use of modern high-yielding varieties bred for intensive, high-input systems. It seems unlikely that increased food production on this scale will come from increasing the area under cultivation. Rather, major solutions will depend upon innovations which exploit sophisticated methods to manipulate plant genomes and the totality of natural genetic variation.

The area of disease resistance will serve to illustrate this argument. Many authors have emphasized the importance of innovation in the area of crop protection since crop losses due to pests and disease may account for between 20 and 40% of productivity worldwide. Indeed, over 50% of the research and development (R&D) spent in the plant breeding/seed industry is focused on the identification and incorporation of resistance traits. In a situation where there is some evidence that the pool of available variation for disease resistance genes within commercial breeding lines is becoming limited (e.g. for yellow rust in cereals) the search is on for novel sources of resistance. There are now many examples of biotechnology being put to the service of pest control, through the incorporation into plants of transgenes for toxins, antifeedants, enzyme inhibitors, etc. But, as Woolhouse (1992) has pointed out, such is the long coevolutionary history of pests and pathogens with their hosts that present strategies for interfering with these relationships represent only a small proportion of the likely defence mechanisms used by plants in resisting foreign organisms. This enormous diversity of mechanisms represents an arsenal of ammunition which can be used in the fight to minimize disease and predation.

Are there any hard facts which support these assertions concerning the value of genetic resources to agricultural productivity? Can we put a cost on the loss of diversity? The sources of variation used by the plant breeder range from current breeding lines to wild species and the products of direct gene manipulation.

In a recent analysis of the plant breeding/seed industry, Swanson (1996) showed that, over a five-year period, 6.5% of all genetic research within this industry, resulting in a marketed innovation, was concerned with germplasm from wild species and landraces.

This compares with only 2.2% of new germplasm arising from technological approaches of induced mutation. The vast majority of germplasm successfully used by the industry still arises from 'conventional' sources, *i.e.* the heavily exploited commercial cultivars, which at first sight might be taken to suggest that wild resources are relatively unimportant in R&D.

In fact Swanson (1996) argues just the opposite, suggesting that the stock of existing commercial varieties should be seen as the information base from which bioindustries develop innovations, whereas the sources of new diversity (wild species, landraces, induced mutation) should be seen as supplying increments to the information base.

In other words, R&D requires an annual injection of 'new' genetic material from natural sources, estimated as amounting to approximately 7% of the stock of material currently within the system.

Biodiversity is valuable to industries other than those that seek novel genes for crop plant improvement. Biodiversity has traditionally provided a source of compounds to the pharmaceutical, foodstuffs, crop protection and other industries. Vast markets exist for the replacement of synthetic biocides, many of which have associated toxicological and environmental problems, by 'natural' compounds, which may have more specific modes of action and fewer side-effects.

Even greater may be the use of genetic resources in more fundamental research where their value in the longer term may be unknown or impossible to estimate accurately at the present time.

Diverse genetic material is needed and now regularly utilized to create mapping populations and to study inheritance of a wide range of traits which may or may not be of immediate practical value. Very large numbers of germplasm samples are regularly screened and evaluated for different characteristics so that the underlying molecular biology can be investigated. The intention here is that a greater understanding of fundamental biology will give rise to wealth creation in the future.

Novel Technologies for Biodiversity Assessment and Monitoring

The analysis and characterization of genetic diversity is fundamental to any *ex situ* or *in situ* conservation strategy. Over the years the methods for detecting genetic diversity have expanded from Mendelian analyses of discrete morphological variants, to biometrical approaches, biochemical techniques based upon protein and isozyme profiles, to methods based on DNA sequence variation.

There is now a huge array of molecular marker techniques, and Chapter reviews the major technologies and their appropriate use in addressing key issues in germplasm conservation The application of molecular technologies to increase our understanding of biodiversity at several levels is exemplified by Chapter, which is concerned with the population level, which is concerned with molecular approaches to genetic relatedness. Chapter reviews how the collection of diverse germplasm for *ex situ* purposes and the development of effective *in situ* collections both require an understanding of the spatial and temporal aspects of population structure and associated genetic diversity and the factors that control it.

Studies with low- or single-copy markers are particularly valuable in answering questions about genetic relatedness between species. Chapter illustrates how comparative molecular mapping provides new opportunities for exploring the genetical basis of phenotypic variation within and between plant species. The implication of comparative mapping is that genes controlling specific phenotypic traits will have corresponding map locations in different taxa. This provides a new framework for the identification of homologous genes in germplasm and new tools for the effective utilization of germplasm in crop plant improvement.

This approach is also valuable in the study of quantitative traits controlled by several genes since flanking markers for major genes provide markers for quantitative trait loci (QTL) manipulation and selection and the map-based cloning of at least some genes controlling quantitative traits.

Biotechnological Tools for Conservation

Biotechnology can make a substantial contribution to the quality of germplasm collections and their efficient management. For example, in the management of very large germplasm collections there is a clear requirement for procedures for fast, reliable taxonomic identification of material and the elimination of redundancy. Samples characterized by molecular criteria may prove superior to traditional analyses of phenotype and examples of the use of DNA markers to guide the management of germplasm are explored in Chapter by reference to one of the world's largest collections of germplasm at the International Rice Research Institute.

Not all plant species can be stored in the form of desiccated seed. Socalled 'recalcitrant' species have to be kept in moist, warm conditions and even then have relatively low longevity. In addition, there are many cultivated plants which must be maintained clonally, such as banana, which produce few seeds, or potato, which does not come true from seed.

To these traditional problem species we must now add the products of biotechnology, new categories of germplasm including clones of elite genotypes, and transformed cell lines, for which storage of seed is inappropriate. Biotechnology has impact on storage technologies in two quite different ways. The first of these, concerns the various types of *in vitro* culture which allow the propagation of plant material in an aseptic environment, thus ensuring the production of disease-free stocks and simplifying quarantine procedures for the international exchange of germplasm. The development of sensitive molecular probes to detect low levels of pathogens.

A more radical approach to safe storage and distribution of germplasm, and possibly the only way to significantly reduce the rate of loss in biodiverse tropical forests, within the bounds of available resources, is to develop alternatives to the costly (and inherently vulnerable) methods of *ex situ* storage of seed or tissue cultures, by conserving germplasm in the form of DNA or DNA-rich materials.

The latter include those dried specimens conserved in the vast resources of the world's herbaria. The DNA Bank-Net, an international network of DNA repositories, has now been established with both storage and distribution functions. Although technologies for rapid removal, analysis, amplification and secure storage of small amounts of DNA are now well advanced, research is still needed on methods to amplify the entire genomic DNA of a species, which means that storage of DNA or DNA-rich materials should still be considered to be something of an insurance policy rather than a replacement for conventional modes of germplasm storage.

Biotechnology-assisted Crop Improvement

Effective use of genetic resources in plant breeding programmes requires methods to determine whether useful genetic variation exists, and costeffective methods to introduce useful genes into the material. Biotechnology has made a contribution to both these aspects.

The use of genetic maps based on molecular markers linked to genes of interest facilitates the more rapid selection of both simple and complex traits thus accelerating their incorporation into breeding materials. Where traits are controlled by one or a few genes, direct isolation of genes and their incorporation into elite cultivars by increasingly efficient transgenic technologies provide an additional approach to the more traditional methods of plant breeding. This prospect also challenges one of the classical concepts in plant genetic resources work, namely the 'gene pool' concept of Harlan and de Wet (1971), which separated cultivated species and their wild relatives into different categories based upon their 'crossability'.

The value of this concept in guiding the work of the plant breeder must now be re-evaluated in the light of molecular genetic technologies which have not only cast a completely new light on the relatedness of crop plant genomes, but which have effectively widened the potential gene pool of a species to include all other life forms.

What challenges does this imply for the traditional plant genetic resources programmes? Why is the emphasis on conserving species that are related to crops when more exotic and potentially valuable genes may be present in liverworts, or green algae, or other less studied taxa?

Bioprospecting for Novel Genes

The value of genetic resources as a source of novel compounds for various industries has been alluded to above. For centuries biodiversity has provided sources of fuels, medicines, foodstuffs, pesticides, etc. The high returns available from introducing a new drug, for example, plus the realization that biodiversity is rapidly being lost (estimated rates of extinction range from 30 to 300 species per day) has encouraged new attitudes to the exploration and exploitation of biodiversity through 'bioprospecting'. How does biotechnology impact on this traditional area? A fundamental operation in any bioprospecting programme is the screening of thousands of extracts for compounds of interest. Gene technology now allows the speeding up of screening programmes for new compounds through the development of more sophisticated *in vitro* assays. For example, the genes encoding receptor proteins for certain classes of drug, or enzymes that may serve as targets for novel pesticides, may be cloned and expressed on a large scale in high-throughput *in vitro* assays into which thousands of plant extracts may be applied.

Bioprospecting is also concerned with discovering novel genes and natural variants of genes or gene products that may either be used directly or to guide sequence improvement by molecular methods. Chapter gives examples of amino acid sequence improvement in peptide hormones as a result of screening homologous genes from the wild, and this area of biotechnology is likely to develop in parallel with new molecular methods of generating random variation through employing the mutational potential of the polymerase chain reaction (PCR) – *i.e.* 'forced molecular evolution' – or recombinatorial libraries of peptides. Whilst gene prospecting has provided a new perspective on biodiversity, the area is not without political controversy because these advances tend to favour agriculture and other industries of developed countries, and may actually displace traditional commodity production in resource-rich but underdeveloped countries. Chapter also outlines the development of strategies which integrate conservation and exploitation with economic returns and sustainable development in those countries from which the bioresources were first obtained.

The Exchange of Information and Germplasm

The value of all plant genetic resources lies in their utilization in crop improvement programmes or in other bioindustries. The vast majority of conserved germplasm is available for distribution to the global scientific community but this international distribution poses considerable risks of accidental introduction of non-indigenous plant pathogens and pests. The detection and elimination of such disease-causing agents is therefore essential.

Often, such agents will be present at very low levels and may actually be symptomless. In this context then, new technologies reviewed in Chapter have been of great value in the development of reliable and sensitive nucleic acid- or antibody-based diagnostic probes, and these will be of great importance in optimizing the health and value of germplasm collections. Finally, if the biotechnology revolution has had great impact in developing novel molecular and cellular technologies, no less a revolution has affected radically the way in which the world organizes and disseminates information.

Powerful computers linked by the Internet have facilitated the development of sophisticated and comprehensive informatics systems which enable information on plant genetics and molecular biology to be cross-related to systematic, ecological and other data through international networks. Chapter explains how this information is organized and distributed against a background of continual development.

Plant Genetic Variation

Effective conservation and use of plant genetic resources involve asking many questions about the extent and distribution of genetic variation. Only when the appropriate markers and technologies for describing this variation are accessible can such questions be adequately addressed.

The most appropriate markers for a given question should:

- Be heritable;
- Discriminate between the individuals, populations, or taxa being examined;
- Be easy to measure and evaluate; and
- Provide results that can be compared with results of similar studies.

Molecular markers and assays that meet these criteria were introduced in the 1950s and have proliferated ever since. Use of protein isozymes as genetic markers increased rapidly after their introduction. Nucleic acid markers gained popularity in the 1970s, with the advent of DNA sequencing and restriction fragment analysis. The number of different molecular markers and assays has increased exponentially since the late 1980s, when introduction of the polymerase chain reaction (PCR) made DNA amplification and sequencing possible and affordable.

As the diversity of markers and assays increases, the cost of using them continues to decrease. At the same time, the range of taxa and tissues that can be analysed has expanded. Improved DNA extraction, amplification, and sequencing protocols allow us to analyse ancient DNA, from 13,000-year-old specimens of *Mylodon* (ground sloths) and 17–20 million-year-old *Taxodium* (bald cypress) specimens, and samples as small as single cells or fractions of a seed.

Thus the chances are good that at least one marker assay will be appropriate for any question about variation in any taxa. Furthermore, results of one study can generally be related to results from many other studies. Such opportunities are inevitably accompanied by challenges. None of the challenges are unique to molecular studies, but all become more critical as the number of available marker assays increases. Faced with a bewildering variety of molecular technologies (and their acronyms), choosing a marker and assay that are appropriate for a scientist's particular question, taxa of interest, and practical constraints can be overwhelming. The temptation is great to use techniques that are recommended by coworkers, are frequently used, for which the most information is available, or are the newest – without adequately evaluating the techniques to determine which is most suitable.

Because molecular markers can be measured in virtually all plant taxa and minimal tissue is required, extra care must be taken to verify that the entries in a test array are correctly identified. Since molecular markers are widely applicable and their constraints are not always understood, thoughtful planning is particularly important when designing molecular studies, to ensure that they have an appropriate range of entries, an appropriate number of entries per taxon, and conclusions that stay within the scope of a specific question and test array.

In addition, the vast amount of published research using molecular markers complicates the task of thoughtfully relating one study to others with similar marker assays, research questions, and/or taxa.

Another challenge is formulating research objectives that ask critical questions about plant genetic resources. Descriptive studies that use molecular marker assays to examine variation between and within specific taxa can generate valuable information, but some plant conservation questions are best addressed by experimental studies. The wide variety of molecular marker assays available provides new ways to address such questions – if the markers are appropriately used. Such challenges can be met successfully and thoughtful decisions can be made if logical frameworks are available to categorize and compare these valuable molecular tools.

A solid foundation for such comparisons has been provided recently. This chapter builds on that foundation by reviewing both early and recently developed molecular markers and assays, citing examples of their use. The types of questions asked about plant genetic resources are described, as are guidelines for selecting appropriate markers and assays to address these questions. The literature cited here is far from comprehensive, but includes applications of each marker assay in plant genetic research. While the list of molecular techniques will soon be dated, hopefully the framework presented will be helpful for categorizing and comparing new tools as they are made available.

The chapter concludes with a perspective on new marker assays as they apply to, and may change, the study of plant genetic variation.

Types of Questions, Markers, and Marker Assays

All studies of genetic variation have four primary components:

- The research question;
- The marker used to address it;
- The technique used to generate and measure the marker; and
- The method used to analyse the data produced. This chapter focuses on the first three components.

Types of Plant Genetic Resource Questions

Most questions asked about *ex situ* and *in situ* plant genetic variation fit into one of three categories. While boundaries between them are not always clear, these categories are useful when deciding how to address specific questions. Questions of identity are forensic in nature and ask whether two or more samples are genetically the same. These questions concern whether geebank accessions are present in duplicate, whether populations are genetically distinct or are geographical subdivisions of one population, and whether genetic change has occurred in an accession or population over time.

In contrast, questions of location and diagnostics concern the presence and location of a particular allele or nucleotide sequence in a taxon, genebank accession, *in situ* population, individual, chromosome, or cloned DNA segment. These questions are asked to locate populations or individuals that have desirable traits; determine whether traits have been lost from a population, or have been transferred from one population or plant to another; and establish the position of markers on physical or genetic maps. The differences between diagnostic and forensic questions are significant, but are not always clarified. Both types of question require qualitative information about specific loci. Yet while diagnostic studies often use single-locus markers, each with few alternative states, forensic questions are commonly addressed with information from many multiallelic loci.

Finally, relationship and structure questions examine relatedness or similarity between genotypes, the amount of genetic variation present, and how variation is distributed between individuals, populations, and taxa. Such questions are used to pinpoint gaps in genebank collections, decide which genotypes are high priority for *in situ* conservation, design germplasm sampling strategies and regeneration programmes, construct core collections, study gene flow, and determine optimum sizes for conserved populations or geebank accessions. These issues are generally addressed with information from many loci. Quantitative data can be used, but qualitative data are more appropriate for some phylogenetic and parentage questions.

Types of Molecular Markers

The appropriate markers for a study can discriminate between entries in an array, but are not so polymorphic that important variation is masked by random noise. Molecular markers range from highly conserved to hypervariable, and can be either proteins or nucleic acids. The nucleic acids used as markers include entire genomes, single chromosomes, fragments of DNA or RNA, and single nucleotides. A wide variety of nucleic acid fragments are utilized as markers. While some occur once in a genome, others are repeated. Many repeated sequences used as markers are noncoding; others are elements of multigene families. Some repeated sequences are interspersed throughout the genome, either distributed randomly or in clusters. These interspersed repeats are common in plant and animal nuclear genomes, and are found in plant (but not animal) mitochondrial genomes. The chloroplast genome contains a large inverted repeat (IR); most angiosperm chloroplasts have two copies, separated by a short single- copy region. Repeat length and (rarely) loss of one copy can vary between taxa.

Much research at present is focused on repeated sequences that occur in tandem. The classes of tandem repeats are distinguished by the length of the core repeat unit, the number of repeat units per locus, and the abundance and distribtion of loci. The names for these classes are themselves varied and have been inconsistently used, but Harding *et al.* (1992) and Tautz (1993) have clarified the nomenclature. Tandem repeats were first reported in the literature as 'satellites' of DNA, detected in CsCl density gradients as fractions with different GC content than the rest of the genome.

These satellites have repeat units that are usually several hundred nucleotides long, with thousands of copies at each of several loci in the nuclear genome. These loci are usually in heterochromatin, often near centromeres. Satellite DNA is present in numerous species. For many satellites, the number of loci and number of repeat units per locus vary between species and higher taxa. Minisatellites (often called variable number of tandem repeat loci, or VNTR loci) are widely used as markers, especially in forensics. The repeat units are usually less than 100 nucleotides long, with tens to hundreds of copies per locus. Thousands of loci in a genome may have similar core repeat units. The number of repeat units at a minisatellite locus can vary greatly between individuals and populations. First described in humans, minisatellites are found in numerous animal species, often near telomeres. They are also common in plants, and are often associated with satellites and centromeres. The number of repeat units per locus is less variable in plants that in animals, but is still high; plant minisatellites are useful markers for variation between and within species.

As suggested by their name, microsatellites – also called simple sequence repeats (SSRs), or simple sequence length polymorphisms (SSLPs) – have very short repeat units, no more than six nucleotides long. SSRs are more abundant than minisatellites in noncoding regions of the nuclear genome, and are present in some nuclear genes and organelle genomes. The number of repeat units per locus is lower for SSRs than for minisatellites, but can approach 100 in animals and 50 in plants. The abundance and polymorphism of SSRs make them particularly valuable for describing variation between populations and individuals. Like minisatellites, SSRs were documented first in humans and later in plants. Plants and animals differ in the abundance of specific SSR motifs in the genome. In both plant and animal genomes, chromosomal distribution of SSRs is variable. Some animal SSRs are found near heterochromatin or interspersed repeats, but most are randomly dispersed. Dinucleotide SSRs are randomly distributed in *Arabidopsis thaliana* (L.) Heynh., but other studies have located plant SSRs near genes, highly methylated DNA, satellites, or centromeres.

Tandemly repeated genes are also utilized as markers. Perhaps the most widely used are the nuclear genes that encode ribosomal RNA (rRNA). The three rRNA genes are separated by two internal transcribed spacer regions, generally referred to as ITS1 and ITS2. These genes and spacers form a unit that is tandemly repeated hundreds of times, at one to several loci in the genome. At each of these loci, the individual repeat units are separated by nontranscribed intergenic spacer (IGS) regions. In the middle region of each IGS are tandem copies of a short subrepeat sequence.

Variation in rRNA gene clusters can be measured at several levels, each evolving at a different rate:

- The number and location of rRNA loci, which is highly conserved;
- The (more variable) number of tandem gene clusters per locus;
- The conserved sequences of the three genes;
- The variable sequences of the ITS regions; and
- The highly variable number of subrepeats in the IGS region.

These features make rRNA gene clusters versatile and informative markers for mapping and phylogenetic analyses.

Types of Molecular Marker Assays

Molecular marker assays are generally classified by whether the molecules evaluated are proteins or nucleic acids, and whether the character analysed in a nucleic acid marker assay is the entire genome, a chromosome, a fragment, or a nucleotide. Alternatively, marker assays can be categorized by the type of character measured. Some methods measure quantitative differences between entries in an array. Others measure qualitative characters, each with two or more possible states. Marker assays also differ in the number of loci evaluated per analysis, whether multiple loci are evaluated simultaneously or sequentially, and the type and amount of information needed about the marker loci before conducting the assay.

Protein Marker Assays: When proteins are used as genetic markers, the assumption is made that any variation between proteins reflects heritable variation in their amino acid sequences. This assumption has long been debated, since protein phenotypes can be affected by factors such as post-translational modification, plant phenology, and environmental conditions during plant growth. For many tasks, however, protein marker assays continue to be useful molecular tools.

Some assays measure immunological properties of proteins. When antigenic proteins of one plant or animal species are injected into an animal (usually a rabbit), antibodies are produced. Genetic differences between entries in an array can be assessed by comparing how each entry's proteins bind to antibody serum (antiserum) from a reference species. This binding affinity is measured by the amount of antibody–antigen precipitate produced or by microcomplement fixation (MCF). Each MCF solution contains antiserum from a reference sample, antigen from a test entry, and a protein complement.

The complement can only bind to antibody that is bound to the antigen. The amount of unfixed complement is measured; this is used to calculate the amount of unbound antigen, which indicates the extent of variation (the immunological distance) between the entry and the reference sample. After MCF assays are conducted for each entry in an array, these immunological distances are compared between entries. MCF has long been used in vertebrate phylogenetic analysis. Until recently, immunological markers were seldom used to study plant variation, perhaps because few suitable proteins had been identified. During the past few years, however, monoclonal antibodies to seed proteins of several cereal species have been produced. Reactivity of these antibodies with antigenic proteins can be evaluated by enzyme-linked immunosorbent assays (ELISAs). Monoclonal antibodies have been used to identify cDNA library clones and describe variation between crop species and cultivars. Other marker assays measure the rate of protein migration through a starch, polyacrylamide, or cellulose acetate gel in response to an electrical current. The migration rate is related to protein size, shape, isoelectric point and/or ionic charge.

In these assays, protein extracts from entries in an array are loaded in adjacent gel lanes and electrophoresed. The gel is then treated with a histochemical stain that reacts with a specific marker protein. Variation between the entries is measured by the positions of their stained proteins on the gel. Many of the proteins used as markers are isozymes – functionally similar forms of an enzyme, encoded by different loci or different alleles at a locus. In general all alleles at a locus are expressed, so allelic isozymes can be analysed as codominant variants.

Isozymes are frequently used to describe variation between and within plant populations, species, and genera; for mapping and diagnostics; and in some cases for identification. When comparing higher taxonomic levels, the probability of genetically different electromorphs migrating to the same gel position is increased. For closely related taxa and populations, however, isozyme analysis is informative and continues to be refined. Immunoelectrophoretic assays measure both protein migration rate in gels and the antibody specificity of proteins. Antigenic proteis are electrophoretically separated, then are detected by their reactions with antibodies. This technique is valuable for diagnostics, mapping, and describing variation between plant cultivars, species, and higher taxa.

Nucleic Acid Marker Assays: Total Genomic DNA and Chromosome Markers: Although proteins are useful as genetic markers, the phenotype each protein describes is determined by the genotype plus the type of tissue sampled, plant phenological stage, environment, and post-translational processing. In addition, the percentage of a genome sampled by protein markers is limited: these markers only describe coding sequences, and not all proteins can be separated and detected by established methods. Some questions are best addressed by markers assumed to be selectively neutral, but this assumption is in question for many proteins. Finally, some proteins are unique to particular taxonomic groups, and thus are suitable as markers only in a limited range of taxa.

In contrast, nucleic acids are present in all organisms and are a common genetic currency for comparing virtually any taxa. Nucleic acid markers describe genotypes, not phenotypes, and can sample both coding and noncoding regions of a genome. DNA methylation and secondary structure can cause artefactual variation in some assays, disrupting the direct relationship between marker and genotype. However, these artefacts can often be detected and eliminated. The first nucleic acid marker assays measured properties of total genomic DNA. When centrifuged in a CsCl density gradient, DNA separates into bands according to GC content. Because highly repetitive (satellite) sequences differ i GC content from the rest of the genome, these gradients can be used to isolate and quantify satellite DNA. This method has been utilized to compare satellite DNA content between plant species, and is still used to detect highly repetitive DNA.

Another technique that separates repetitive and low-copy DNA relies on DNA reassociation kinetics. The two complementary strands of a DNA molecule denature when heated and reassociate into duplexes when cooled. At a standard temperature and salt concentration, reassociation rate depends on the percentage of strands with similar sequences. Repetitive sequences reassociate rapidly and at high temperatures, compared to single-copy sequences. Thus highly repetitive, middle repetitive, and single-copy fractions of a genome can be isolated by fragmenting, heating, and cooling double-stranded DNA. Because reassociation rate is affected by fragment length and the pattern of sequence repetition, the type and dispersion of repeats can be measured by varying the lengths of the fragments analysed. This method has been used to characterize and compare repeat sequences between plant species and hybrids.

DNA–DNAhybridization employs the same principles, but with single-copy sequences, to compare entries in an array. When single-stranded DNA from two entries is combined, their overall sequence similarity is related to the temperature at which 50% of the strands form heteroduplexes. This *T*m is compared to the homoduplex *T*m for each entry; D*T*m measures the genetic distance between entries. DNA–DNAhybridization is used to examine variation between species and higher taxa, usually with vertebrates and less often with plants and bacteria. Variation in DNA content of nuclei and chromosomes can be measured by flow cytometry. Particles are isolated, suspended, stained with a DNA-binding fluorochrome, and passed through a cytometer. The fluorescent signal intensity from each particle indicates its DNA content. Flow cytometry was first utilized to analyse mammalian genomes, but has since been used to compare nuclear DNA content and ploidy level between plant families, genera, species, and ecotypes and to evaluate widehybrids. Becaus chrmosomes can be identified by their signal intensity, flow cytometry can be used to generate high-resolution karyotypes and sort chromosomes for mapping.

Karyotypes of chromosomes prepared in mitotic metaphase squashes have long been used for diagnostics, mapping, and describing genetic variation between plant species and hybrids. Entries in an array can be characterized by their chromosome number and morphology. In addition, chromosome spread preparations can be treated with stains specific to AT-rich regions, heterochromatin, or nucleolar organizing regions. The stain-specific banding patterns of individual chromosomes are then used as markers to characterize and compare entries.

Chromosome banding is informative for mapping studies and describing variation between geneban accessions and cultivars. Banding is often used in combination with flow cytometry or other techniques. Fo chromosomes that are very small or lack distinctive bands, cytogenetic techniques may have limited use. As high-resolution microscopes and more sensitive stains are developed, however, the value of these assays continues to increase.

Nucleic Acid Marker Assays:Fragment Markers: Nucleic acid fragments are widely used as genetic markers. As mentioned above, CsCl density gradients and DNA reassociation techniques can detect and roughly quantify satellite DNA sequences. Yet these methos do not measure discrete character differences or variation in the sequences themselves. In addition, differences in GC content between genomes can significantly affect the results. In contrast, fragment markers provide high-resolution qualitative information about sequence variation. Either DNA or RNA can be evaluated, and minimal amounts are required. Numerous fragment marker assays are available; all involve detecting fragments that differ in presence, size, or quantity between entries in an array.

Fragments that are produced, electrophoresed, then detected: In many assays, fragments are produced, separated by agarose or polyacrylamide gel electrophoresis, then detected. The length and conformation of a fragment determine its migration rate in a gel. Several different approaches are taken for detecting fragments. In some assays with purified sequences, all fragments are detected and visualized. The fragments are either labelled before or stained after electrophoresis. Ethidium bromide, silver, fluorescent, and radioactive stains and labels are commonly used. In other assays, specific target fragments are detected and visualized by transfer hybridization. All fragments in an assay are electrophoresed, denatured if double-stranded, then transferred from the gel to a nylon or nitrocellulose membrane. A labelled single-stranded probe complementing the target sequence is hybridized to the membrane and anneals to the target fragments. The extent of mismatch hybridization varies with salt and formamide concentration and the temperature during hybridization. *A variety of probes can be used*: Synthesized oligonucleotides, uncharacterized clones from genomic DNA libraries, fragments isolated from other gels, RNA sequences, or total genomic DNA. The probe may be from the same species as the membrane-bound DNA or from a different species. Until recently most probes were radioactively labelled, but nonradioactive colourimetric, fluorescent, and chemiluminescent labels are increasing in popularity).

Fragments produced by restriction digest: Restriction site analysis is widely used in plant genetics. The principles, protocols and data interpretation have been discussed in detail. Restriction fragments are generated by treating double- stranded DNA with restriction endonucleases (REs), which are enzymes produced by bacteria. Each RE cleaves double-stranded DNA at a specific sequence, usually four to six nucleotides in length. Several hundred REs have been characterized; they vary in recognition sequence and ability to digest methylated DNA. DNA sequence variation between the entries in an array is detected as variation in the number and/or lengths of restriction fragments. This variation can be caused by point mutations at RE recognition sites or by the insertion, deletion or rearrangement of sequences at or between RE sites. When single- or low-copy fragments are evaluated, variations in fragment length are referred to as restriction fragment length polymorphisms (RFLPs). In some RFLP assays of highly purified DNA sequences, all of the fragments are visualized by labelling them before or staining them after electrophoresis. In most RFLP assays, however, target fragments are detected by hybridized probes. For each entry in an array, the presence or absence of each individual fragment is recorded.

These fragment data are used to infer the presence or absence of each restriction site. For diploid genomes, entries heterozygous for the presence of a restriction site can be distinguished from those that are homozygous. Although few RFLP loci are examined on each gel, every locus can be genetically defined. RFLP analysis can be expensive, but costs per trial can be reduced by probing each membrane several times with different probes. Single- or low-copy RFLPs are used extensively in diagnostics, for mapping, and to verify interspecific hybridization and study genetic relationships and structure. Probes that hybridize across species are especially useful for comparative mapping.

In most restriction site analyses of repetitive DNA, target fragments are detected by probing with a known repeat sequence. The fragments, not the genetically defined sites, are usually the characters analysed. Entries that are homozygous for the presence of a fragment cannot be distinguished from heterozygotes. These assays provide information about variation at many loci simultaneously, but reveal little about the type of sequence variation. Restriction site analysis is often used to study interspersed repeats and satellites. It was utilized in the first assays of hypervariable loci, and is presently the most common method in forensics for evaluating minisatellite VNTR loci. Multilocus restriction fragment profiling of plant genomes is useful for identifying clonal cultivars and studying breeding systems. In some studies, fragment length polymorphism can be interpreted as variation of codominant alleles at a locus.

However, because inheritance of markers in multilocus profiles is seldom clearly defined, their use in phylogenetic and gene flow studies may be limited.

Restriction site analysis can generate highly reproducible results, if all steps of the process are carefully controlled. With a variety of probe– enzyme combinations available, assays of the nuclear genome can be used to address many questions about genetic variation at or below the species level. As in isozyme analysis, the probability of genetically different restriction fragments co-migrating on a gel is increased when higher taxonomic levels are compared. Because chloroplast genomes evolve slowly, chloroplast restriction site assays are often used to study variation between higher taxa. In some species, both chloroplast and nuclear genomes are biparentally inherited. In many plant species, however, chloroplasts are either maternally or paternally inherited. For these species, chloroplast restriction site analysis is very informative for studies of gene flow and interspecific hybridization, especially when used in combination with assays of biparentally inherited nuclear markers.

Fragments produced by the polymerase chain reaction: Fragment analysis began in the 1970s with the introduction of RFLPs, and grew exponentially a decade later when the polymerase chain reaction (PCR) was developed. PCR uses the principles of DNA reassociation and the action of DNA polymerase to amplify nucleic acid fragments *in vitro*. Each PCR reaction solution contains a double-stranded DNA template, short single-stranded oligonucleotide primers, thermostable DNA polymerase, enzyme cofactors, and the four deoxynucleotides (dNTPs: dATP, dCTP, dGTP, dTTP).

The template is denatured by heating; the temperature is lowered, and the primers anneal to complementary sequences on the template. The solution is heated again, and the polymerase adds dNTPs to the 3' end of each annealed primer. Double-stranded fragments are produced and serve as templates for the next cycle, which generates fragments with primer sequences at each end. During subsequent cycles, these fragments are amplified at a geometric rate.

With automated thermocyclers and cheaper, improved reagents, PCR is fast becoming more reproducible and affordable. A wide range of known and unknown sequences can be amplified from fresh or ancient DNA and from RNA, and little template is needed; hence PCR is applicable to any taxon. However, contamination is a concern, since contaminant DNA in a reaction solution can function as a template. The specificity and reproducibility of PCR depend on several factors: the concentration and quality of the reaction ingredients; the design and GC content of the primers; and the temperature, duration and number of cycles. However, measures can be taken to minimize and detect contamination and optimize cycling protocols. The disadvantages of PCR are greatly outweighed by the tremendous opportunities it provides. When amplified fragments are used for high-resolution mapping, diagnostics or forensics, they are generally detected by probes. When used to describe variation between and within species, amplified fragments may be detected by labelling primers or dNTPs in the reaction solution, labelling PCR products, or staining gels after electrophoresis.

Numerous PCR-based marker assays have been developed and used in plant genetics. They can be categorized by whether target sequences are known prior to amplification, whether the primer sequences are designed or are arbitrary, the number of primers, and the size range of amplified products. In the first PCR assays the target loci were known singlecopy sequences, amplified by two primers designed to complement the regions flanking the loci. Primer pairs have since been designed for numerous single-copy loci in plant and animal nuclear and organelle genomes. These loci are very useful as markers for mapping, diagnostics, phylogenetic analyses, and studying hybridization.

In most genome mapping projects, cloned DNA sequences present at only one site in a genome are used as physical mapping landmarks. Primer pairs have been designed to amplify many of these landmarks as sequence tagged sites (STSs). This strategy is used extensively in the Human Genome Project (HGP) and other mapping programmes. In assays of these designed-primer PCR loci, the character evaluated may be the fragment itself, with presence dominant over absence. Often, however, inheritance can be defined.

For loci in diploid genomes, differences in fragment length can then be analysed as codominant alleles. Both interspersed and tandemly repeated sequences can be amplified by designed-primer PCR. When several interspersed repeat loci are amplified in one sample, each locus is evaluated separately. When tandem repeat loci are amplified, length polymorphism at a locus is assumed to reflect variation in the number of tandem repeat units.

Amplification of human minisatellites with designed primer pairs was soon followed by amplification of plant minisatellites and SSRs. Amplified plant SSRs are informative for mapping, genotyping, diagnostics, and population studies. SSR alleles may differ little in length, and discriminating them can be difficult. New electrophoresis, labelling and imaging technologies are making detection easier.

Development of designed-primer PCR markers involves several steps. Target sequences are identified and isolated, usually from a genomic DNA library; the loci and flanking regions are sequenced; and primer pairs are designed and synthesized. Currently these tasks are expensive, and not all laboratories have the equipment required. Once primers are designed and synthesized, however, the costs for each analysis are not high. The initial expenses of marker development may be reduced by identifying the desired sequences from nucleic acid databases, *e.g.* Genbank or EMBL. However, this is only applicable to species well represented in the databases. As another cost-cutting measure, some primers designed to amplify loci in one species can be used to amplify loci in related taxa. Primers that amplify across taxa are also useful for phylogenetic analyses and studying genome evolution.

For development of designed-primer PCR markers, both target and flanking region sequences must be known. Soon after PCR was introduced, however, its ability to amplify unknown sequences was realized. In inter-repeat PCR, primers that complement interspersed repeats or other known sequences are used to amplify unknown sequences between the repeats. 'Universal' primers have been designed to complement conserved coding sequences in mitochondrial and chloroplast genomes; these primers can be used to amplify the noncoding regions between genes.

Other marker assays require no prior sequence information, for either the target region or the flanking regions. These assays – randomly amplified polymorphic DNA analysis (RAPD), arbitrarily primed PCR (AP-PCR; Welsh and McClelland, 1990), and DNA amplification fingerprinting (DAF) – use one short primer that has an arbitrary sequence. This primer anneals to the template at complementary sequences in both 'sense' and 'antisense' orientation, and the regions between primers in opposite orientation are amplified. The number and length of fragments amplified by one primer can vary; short primers generally produce a large number of small fragments. In single-primer PCR, entries that are homozygous for the presence of a fragment cannot be distinguished from heterozygotes; thus fragment presence is generally dominant over absence. Many loci can be evaluated simultaneously, although little genetic information about each locus is provided. Arbitrary PCR is particularly valuable when no sequence information is available for the taxa being studied. It has been used extensively to examine variation between and within plant species, identify genebank accessions and cultivars, and construct genetic maps.

Many PCR assays are variations or combinations of designed-primer PCR and arbitrary PCR. Most were developed to increase the reproducibility of results and/or to better discriminate between closely related genotypes. In arbitrary PCR, reproducibility and the extent of mismatch primer annealing are highly dependent on the cycling parameters and concentration of reaction components. One solution to this problem is to identify polymorphic fragments in arbitrary PCR assays and then isolate, clone, sequence and design primers for these fragments. This process generates new designed-primer PCR loci but involves many steps. Other assays with increased reproducibility combine features of single-primer and interrepeat PCR: unknown sequences are amplified by one primer that complements an SSR, its flanking region, or both. Finally, 'anchored' PCR uses one designed and one arbitrary primer to amplify unknown sequences.

For genotyping, diagnostics, and assessing variation between close relatives, PCR-based marker assays with high resolution are needed. One such variation is nested PCR. In this technique, one amplification reaction with two designed primers is followed by a second reaction. The first PCR product is the template for the second reaction, which uses two designed primers internal to the first set. Although the possibility of contamination is high, nested PCR is valuable for extremely sensitive genotyping and diagnostics. At present, it is used most often in medical research.

Resolution of arbitrary PCR can be increased by adding stable mini-hairpin sequences to the 5' ends of short primers. This technique greatly improved discrimination between cultivars of centipedegrass [*Eremochloa ophiuroides* (Munro) Hackel]. Other marker assays with improved resolution combine principles of PCR and restriction site analysis. RE digestion of PCR templates before they are amplified was first suggested by Mullis and Faloona (1987). Arbitrary PCR with cleaved templates has since been used to map plant genomes and identify genotypes. Amplified fragment length polymorphism (AFLP) analysis is a variation of cleaved-template PCR. After DNA is digested, a short adaptor sequence is ligated to each end of the restriction fragments. The fragments are then used as templates for single-primer PCR. The primer contains a constant sequence (the adaptor and RE recognition site sequences) plus selective bases that determine the specificity of primer annealing. Both reproducibility and resolution are increased by using AFLPs. Resolution can also be increased by digesting PCR products before electrophoresis. Digested products of both designed primer PCR and arbitrary PCR have been used to measure variation between genebank accessions, refine RFLP maps, and study interspecific gene flow. Resolution of both PCR and restriction site assays can be improved by denaturing double-stranded DNA fragments before they are electrophoresed. Due to internal point mutations or differences in conformation or GC content, the two single strands from one fragment may have different migration rates on polyacrylamide gels. Both denaturing gradient gel electrophoresis (DGGE) and single-stranded conformation polymorphism (SSCP) analysis can detect differences between closely related genotypes.

Fragments detected directly by probe hybridization. In some marker assays the target sequences are not separated by electrophoresis, but are detected directly by probe hybridization. The types of probes used are the same as in electrophoretic assays, and can be radioactively or nonradioactively labelled. In dot and slot blot hybridization, single-stranded DNA fragments, genomic DNA, or RNA is immobilized on a nylon or nitrocellulose membrane. Labelled probes complementing the target sequence are hybridized to the membrane blot. All loci in the blot are sampled simultaneously; abundance of the target sequence is measured by the intensity of the hybridization signal. Many blots on one membrane can be evaluated at the same time, and the procedure is relatively simple. Because the presence of nontarget sequences in the blot can affect hyridization, precise measurment of target sequence abundance in genomic DNA blots can be difficult. When the blots are carefully prepared and hybridization conditions are controlled, however, these assays can generate valuable information for diagnostic and relationship studies. Dot and slot blots have been used to detect specific alleles of genes and RAPD loci, estimate the copy number of satellites and interspersed repeats, and compare gene expression between organs of a plant.

In another technique, *in situ* hybridization (ISH), probes are hybridized to the same chromosome spread preparations used for karyotypes and banding pattern assays. ISH is used to detect the presence and chromosomal distribution of target sequences. While banding assays are only appropriate with chromosomes that contain specific sequences, ISH can detect a wide range of sequences. Radioactive and nonradioactive probes are used, but the popularity of fluorescent *in situ* hybridization (FISH) is rapidly increasing.

ISH has many applications for genome analysis. These include 'painting' the genome of one species onto chromosomes of another, to study species evolution; painting a genome with one of its own chromosomes, to study chromosome evolution; and characterizing the chromosomal distribution of interspersed repeats. ISH is often used in combination with other marker assays. In wide crosses and somatic hybrids, ISH with species-specific interspersed repeat probes can indicate the abundance and chromosomal locations of transferred sequences; restriction site analysis can provide information about structure and rearrangement of the sequences. Physical mapping with ISH can complement RFLP genetic mapping studies.

Aledo *et al.* used slot blots to estimate the copy number of an interspersed repeat in the maize (*Zea mays* L.)genome, and used FISH to determine the chromosomal distribution of the repeat. With both ISH and heterochromatin-specific staining on the same chromosome preparation, Harrison and Heslop-Harrison (1995) confirmed that two *Brassica* satellite families are located near centromeres. By varying the hybridization stringency, differences between satellite sequences on different chromosomes were revealed.

ISH can be time consuming, and results can be affected by the specific methods used for chromosome preparation, denaturation, and probehy-bridization and detection. However, new protocols are increasing the reproducibility of ISH.Fukui *et al.* (1994) used a thermocycler to control temperatures during denaturation, and developed accurate imaging methods for detecting and analysing fluorescent signals. Because of such innovations, the value of ISH as a molecular tool will most likely continue to increase.

Nucleic Acid Marker Assays: Nucleotide Sequencing: Nucleic acid variation can be described most directly by using the nucleotides themselves as discrete characters, each with four possible states. Because nucleotides are the basic units of genetic information in all organisms, sequencing can be used to detect very low levels of variation. Nuclear and organelle genomes of virtually any taxon can be sequenced; minimal amounts of DNA or RNA are needed, and fresh or preserved DNA can be used. With most marker assays, entries in a trial are compared to each other; results are not easily compared across trials that have no standard entries in common. In contrast, sequences from one study can be compared with sequences from any other study. In addition, sequencing can reveal whether variation is due to nucleotide substitution or sequence rearrangement.

Two sequencing methods have been used since the late 1970s, and are described in detail by Hillis *et al.* (1990).

In both methods, labelled fragments are generated from DNA or RNA that has been isolated and purified; the fragments are electrophoresed; and the sequence is 'read' from the fragment patterns. Maxam–Gilbert (chemical) sequencing is used to sequence double-stranded DNA. DNA is endlabelled, then divided into four reaction solutions.

Each solution contains one of four reagents, each of which cleaves DNA at specific bases. The sequence of the DNA determines the lengths of the fragments produced. *In vitro* (dideoxy chain termination) sequencing is applicable to both DNA and RNA. The single-stranded template is divided into four reaction solutions. Each solution also contains the following: a labelled oligonucleotide primer that complements one end of the template; the four dNTPs; and one of four dideoxynucleotides, each of which terminates extension at one of the four bases. In each reaction, the primer is extended along the template. As in chemical sequencing, the template sequence determines the lengths of the fragments produced. Chemical sequencing is preferred for some genomic regions with complex secondary structure; in general, however, the chain-termination method is more commonly used. Since it has been coupled with PCR, combined with fluorescent labelling and automation, this technique is rapidly becoming the method of choice.

Sequencing can be used to address a wide range of plant genetic resource questions about virtually any taxa, but involves considerable investment of money, labour and time. Automation and large-scale production of cheaper reagents have made sequencing more cost effective, but few laboratories can afford to sequence large genome segments from multiple entries. As a result, in many studies only one locus or a few loci are sequenced for a limited test array. While many adjacent nucleotides may be evaluated, together they sample only one locus. Thus sequencing is usually not practical for mapping, multilocus forensics, population genetics, or analyses of intraspecific variation. It is often used for high-resolution diagnostics, studying specific genes, and describing variation between genera and higher taxa. In phylogenetic studies, however, inferring genome evolution from sequence variation at a single locus should be done with caution if at all. Many regions of plant nuclear and organelle genomes have been sequenced and used as markers. These sequences vary greatly in evolutionary rate and the taxonomic levels at which they are employed. Chloroplast genomes evolve slowly; sequences of the entire genome or its *rbc*L gene are informative as markers for describing variation between genera and higher taxa. Nuclear genes that have been sequenced and used as markers include the phytochrome, heat shock protein, and nuclear rRNA gene families. In general, the rRNA coding regions are highly conserved; internal transcribed spacers are more variable, and intergenic spacers are highly polymorphic. Finally, sequencing is often used – in combination with other marker assays – to characterize tandem repeat loci.

While sequencing may be the ultimate molecular tool, caution should be taken to avoid several potential problems. Because each nucleotide is a separate character, sequence fidelity is crucial. Thus nucleic acids must be carefully isolated and purified before they are sequenced. In addition, PCR contamination and amplification errors have more serious effects for sequencing than for fragment analysis. Amplification errors can be minimized by optimizing PCR conditions and using polymerases with proofreading capability. Mistakes during gel reading (manual or automated) must also be considered and minimized. Subjectivity is a real concern when aligning sequences before they are compared. Alignment is complicated and requires assumptions about expected amounts of base substitution and insertion/deletion. Despite these concerns, sequencing remains the tool of choice for many studies.

Guidelines for Selecting Marker Assays

Choosing appropriate marker assays can be challenging, but several considerations can make the task easier.

Important issues are:

- What question is being asked?
- What level of resolution is required?
- How can the results be related to characteristics of the taxa being studied?

Questions of identity are best addressed with high-resolution assays and large numbers of codominant marker loci that are evaluated sequentially. Electrophoretic assays of isozymes and nucleic acid fragments meet most of these criteria. Genetically defined isozymes, amplified SSR loci, and single-copy RFLPs are very informative markers for identification. Due to practical constraints, however, multilocus assays of restriction fragment and PCR markers (VNTRs, RAPDs and AFLPs) are often used. Dot blots, sequencing and monoclonal antibody assays are also appropriate, but the costs of analysing many loci per assay may be prohibitive.

The most appropriate assays for diagnostic questions have high levels of resolution and use genetically defined, codominant markers. Because these questions can often be addressed with one or few loci, useful methods include electrophoretic assays of isozymes, single-copy RFLPs, or SSRs; dot or slot blot assays; and sequencing.

For some questions, monoclonal antibody assays and ISH are also appropriate. Mapping studies often use a combination of tools: isozyme or DNA fragment electrophoresis for genetic maps, and chromosome banding, flow cytometry, ISH, and STS markers for physical mapping.

For questions of relationship and structure, important issues include the type of question asked and the range and levels of taxa evaluated. Because evolutionary questions must be answered with qualitative data, the assays used for diagnostics and identification are appropriate. Questions of similarity can be addressed with these qualitative marker assays or with quantitative assays (microcomplement fixation, flow cytometry, CsCl density gradients, and DNA–DNA hybridization).

The latter two techniques have low levels of resolution, and are generally used to compare genera and higher taxa. Interspecific and intraspecific variation is more appropriately described with isozyme and DNA fragment markers, which have higher levels of resolution. In electrophoretic assays of isozymes and DNA fragments, co-migration of genetically dissimilar electromorphs or fragments is likely when higher taxa are compared. If resolution is too high, however, the assays may be numerically precise but contain random 'evolutionary noise' that masks meaningful variation. Researchers can often decide whether a particular marker assay is suitable for a given study by reviewing the results of previous studies using that technique.

Several characteristics of the taxa being studied are important when selecting a marker and marker assay. One of these characteristics is the mating system. Many apomictic and self-pollinated plant species contain little variation, so techniques with high resolution may be required. When deciding whether to use a chloroplast genome marker, information about chloroplast inheritance is critical.

For many marker assays, polyploidy can be a concern. When polyploid taxa are evaluated, the number of loci sampled, the problem of dissimilar fragments co-migrating, and the level of data complexity are often increased. This complexity may be reduced by using genetically defined isozymes, chloroplast genes, genes that evolve slowly, or single-copy markers – rather than RAPDs or multilocus VNTR profiles.

Practical considerations are often the most important. The amount and quality of proteins or nucleic acids required can differ greatly between marker assays. Sequencing and PCR-based assays require little template and can be used to analyse ancient DNA samples, but are very sensitive to contamination. DNA–DNA hybridization requires large amounts of DNA, but small amounts of contaminant DNA are less of a problem. For some species, fresh tissue is required for isozyme electrophoresis. For some species and marker techniques, the presence of secondary compounds can complicate storage and extraction of DNA or protein, and can interfere with marker detection.

Marker assays also vary in the amount of setup work required and the extent of sequence information needed before conducting the assay. Karyotyping, ISH, and RFLP assays involve preparing chromosome spreads and/or probes. In contrast, some PCR-based assays are streamlined for high throughput and require little setup work. Sequence information is needed for probe hybridization and designed-primer PCR assays, but not for arbitrary PCR. Finally, the resources available must be considered. Minimal equipment is needed for protein electrophoresis, but other techniques require thermocyclers, fluorescent microscopes, cytometers and radioactive labelling.

Choices between newly developed and well-documented techniques should be made in light of the expertise and personnel available. Other considerations are reproducibility, number of entries and loci sampled per analysis, and cost per unit of information. Comparative studies can make selection of a marker assay easier and reveal where costs can be reduced. Costs and technologies are constantly changing, however, and the comparisons themselves may be expensive and time consuming; their value must be weighed against the resources they require and the necessity of continual updates. When comparative studies are not feasible, thoughtful consideration of the issues described here may be helpful.

Bibliography

Altered Fates: *Gene Therapy and the Retooling of Human Life*, W. W. Norton and Company, New York, 1995.

Bellairs, A.: *The Life of Reptiles*, Weidenfield and Nicholson, London, 1969.

Curan, Brendan: *A Terrible Beauty is Born: Clones, Genes and the Future of Mankind.* New York: Taylor & Francis, 2003.

Dhawan, Ruchi: *Cloning, Genetics and Stem Cell Technology : Development and Applications*, Axis Pub, Delhi, 2010.

Fisher, R. A.: *The Genetical Theory of Natural Selection*, Clarendon Press, Oxford, 1930.

Friedberg, E.C.: *DNA Repair*, New York, WH Freeman and Company, 1985.

Gall, Joseph G.: *Landmark Papers in Cell Biology,* Cold Spring Harbor Laboratory Press, Plainview, NY, 2001.

Glover, D. M., and Hames, B. D.: *Genes and Embryos*, Oxford University Press, New York, 1989.

Hartl, Daniel L.: *Basic Genetics*, Jones and Bartlett Publishers, Boston, 1991.

Jones, Gareth D., and Mary Byrne: *Stem Cell Research and Cloning: Contemporary Challenges to Our Humanity*. Hindmarsh, SA, Australia: Australasian Theological Forum, 2005.

Kass, Leon R and James Q Wilson: *The Ethics of Human Cloning*, Scientia, Delhi, 2002.

Kumar, Rajesh: *Fundamental of Artificial Neural Network and Fuzzy Logic*, Laxmi Pub, Delhi, 2009.

Lamba, V.K.: *Neuro Fuzzy Systems*, Laxmi Publications, Delhi, 2008.

Lawrence, Jeanette: *Introduction to Neural Networks*, California Scientific Software Press, 1994.

Margulis, L.: *Symbiosis in Cell Evolution,* W.H. Freeman, San Francisco, 1981.

Roetz, Heiner: *Cross-cultural Issues in Bioethics: The Example of Human Cloning.* Amsterdam: Rodopi, 2006.

Ross T : *Fuzzy Logic with Engineering Applications.* McGraw-Hill, New York, 1995.

Scriver, C.R.: *The Metabolic and Molecular Basis of Inherited Disease,* McGraw Hill, New York, 2001.

Silver, Lee M: *Remaking Eden: How Genetic Engineering and Cloning Will Transform the American Family.* New York: Avon, 1998.

Stoll RR : *Set Theory and Logic,* Dover Publications, New York, 1979.

Sugeno, M.: *Industrial Applications of Fuzzy Control,* Elsevier Science Pub. Co., 1985.

Von Altrock, Constantin: *Fuzzy Logic and NeuroFuzzy Applications Explained.* Upper Saddle River, NJ: Prentice Hall, 1995.

Wilmut, Ian, and Roger Highfield: *After Dolly: The Uses and Misuses of Human Cloning.* New York: W. W. Norton, 2006.

Wilson, E. O., and Lumden, C.: *Genes, Mind, and Culture: The Evolutionary Process,* Harvard University Press, Cambridge, Mass, 1981.

Woodward, John: *The Ethics of Human Cloning.* San Diego: Greenhaven Press, 2004.

Wright, L.: *Twins: Genes, Environment and the Mystery of Human Identity,* Weidenfeld & Nicolson, London, 1997.

Index

❑❑❑